T0276177

THE EVOLUTIONARY COSMOS:

OUTSIDE-IN THINKING THE UNIVERSE

Richard Westberg and Cal Orey

Cover Graphic: A. Khalatyan/J. Fohlmeister/AIP An artist's depiction of what the scientific data is suggesting is the biggest structure in the Universe, some of which could be an exterior invisible, magnetic double helix that is not shown. The double helix could be herding the galaxies, most likely in a straight line, as it rotates.

Special Feature: Links will work on ebook devices.

First edition, 2022

AuthorHouse™
1663 Liberty Drive
Bloomington, IN 47403
www.authorhouse.com
Phone: 833-262-8899

This book is printed on acid-free paper.

ISBN: 978-1-6655-5471-8 (sc)
ISBN: 978-1-6655-5470-1 (hc)
ISBN: 978-1-6655-5472-5 (e)

Library of Congress Control Number: 2022905105

Print information available on the last page.

Published by AuthorHouse 04/05/2022

authorHOUSE®

A Special Dedication . . .

This book is dedicated to my son Jack, who always believes in me, no matter what crazy curveballs the cosmos tosses my way. He said I was his hero, which made me proud. All others in my life were skeptical about my innovative and imaginative thinking, while he never was. I aim to make a difference for people who believe there are unsolved puzzle parts to the beginning, present, and future of the Universe as we know it.

Table of Contents

Part 1: IN THE BEGINNING **1**
1. The Creation of "Shaylik" 2
2. Chiral-Shaylik Connection 9

Part 2: SHAYLIK AND PLANETS **15**
3. Dust Storms on Mars 16
4. Ice Volcanoes at Ceres 21

Part 3: THE UNIVERSE IN A NEW LIGHT **25**
5. A Modern Meaning of Gravity 26
6. Discovering Dimensions to Anyons 31
7. More Surprising Shaylik Observations 38
8. A New Look at X-Rays 42

Part 4: BUILDING BLOCKS TO A SOLAR SYSTEM **47**
9. Earth's Magnetic Force 48
10. Say Hello to Planet Mercury 53
11. Solar Superstorms 64

Part 5: MORE MATTERS OF SHAYLIK **69**
12. Electrifying the Earth 70
13. Flow of Shaylik 82
14. Fractal Explained 90

Part 6: AS THE WORLD'S THEORIES TURNS **95**
15. Darwin's Theory of Evolution 96
16. Rethinking Einstein's Gravity Theory 99
17. The Evolution of Our Universe 103

A FINAL WORD 107
The Cosmos Glossary 113
NOTES 115
Epilogue 118

PART 1

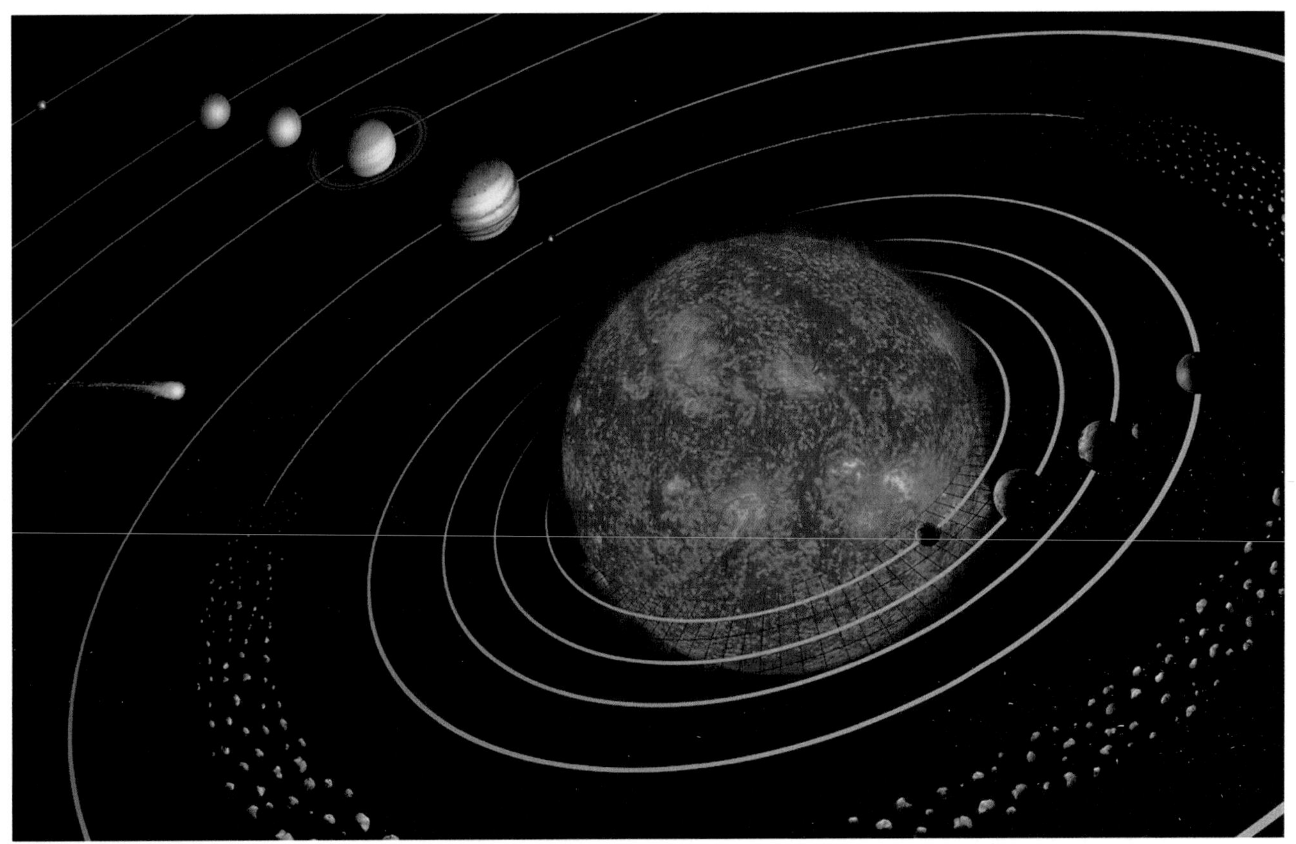

IN THE BEGINNING

CHAPTER 1

The Creation of "Shaylik"

What we know is a drop, what we don't know is an ocean.
—Isaac Newton

Isaac Newton was right. The information I discovered during my quest online discusses the current problems at length. The first article says that our scientists have themselves in a technological corner, which has, as yet, no way out. What they need is unconventional thinking, which is what this book is. This is why I believed this book needed to be written. (1, 2)

Several years ago, my son Jack and I packed a few bags, filled up my vehicle, stocked up on food, and left our home for another nature adventure to California. We set out on a traditional annual summer fishing trip to Lake Sabrina, California, southwest of Bishop, and nestled in the Inyo National Forest.

In the two-day road trip, we had plenty of idle time to talk about meaningful topics, such as life and our Universe (everything including infinite space and time). It beats going to a man-made luxury hotel or flying to a tourist hot spot with crowds. And naturally, discussing notions about the ever-changing Outer space nourishes inner brain and soul.

I recall one particular journey back in 2014. At night under a Full Moon, we sat around a warm campfire. Stargazing on a clear night with a view of the Milky Way and the Hercules Globular Cluster, we sipped cold beer and chewed on stuff about Outer space. Then I brought up the subject of gravity (the force that pulls what everything is made of together)—you know, the topic attributed to Newton and the apple tree, or the phenomenon described by Albert Einstein's theory of general relativity, known as "the result of massive objects curving the fabric of space-time."

Truth be told, gravity is a subject that has been on my mind since I was a teenager. While science was my favorite subject in high school, I never could accept my teachers' clarifications when I listened to them talk about gravity. None of the explanations and details seemed logical. All I knew was that I wanted to know more and was left with less, but it didn't stop my fascination with the Universe (everything including infinite space and time).

As time passed, I realized the most glaring things we don't know about are Dark Matter ("a form of matter that only interacts with 'normal' matter through gravity") and Dark Energy ("a form of energy that permeates the Universe, making up a percentage of the Universe's total mass-energy content; and has an antigravity effect"), which science gurus admitted was 80 to 90 percent of the explanation for why galaxies stay together. This expanded to the admission that the "experts" are unaware of 90 percent of the mass (a measurement of how much matter is in an object) that exists in the Universe. And I was left dazed and confused.

It made me think I was left on my own without a definitive answer. If I were talking to an investment advisor, for instance, because I wanted to invest my life savings in a safe place and the consultant told me that he did not understand 90 percent of why the stock market works the way it does, I would have run, not walked, out of the premises. And I'd be left in the dark with questions left unanswered.

A Fishing Hole for Answers to Outer Space

The cabin we stayed in was 20 minutes away from Lake Sabrina, California. The region had been a mining camp before it was converted to a rustic fishing and hunting site. As luck would have it, it was also used as a religious retreat—a sign for why we'd talk about the Heavens, perhaps.

The cabin we stayed in was the only two-story cabin of about 12. We never rented the upper part. The ground floor had two bedrooms with four bunk beds each and a kitchen with a stove, refrigerator, sink, and large dining table for eight. There was also a spacious living room with one couch and multiple chairs. The log cabin was an ideal place to contemplate the Universe.

An old gold mine was a 30-minute walk away from a bridge that went over the small river from Lake Sabrina. There was a small pond five minutes from the cabin. It always had trout in it.

The weather was cold at night due to the high altitude. We had everything from snow and lightning storms to 80-degree days. Each year was different. There were aspen trees everywhere as well as towering pines. Our view of the sky was breathtaking but limited by the tall mountains around us. The setting was the ideal place to ponder the cosmos. The philosopher Pythagoras first used the term "cosmos" (Ancient Greek) for the order of the Universe. (3)

Over the years. the attendees were anywhere from 20 to late seventies. We always arrived on Sunday and left the following Saturday. We even went back in 1999, the end of the decade when people were concerned about the effects of Y2K. (This refers to the millennium bug that could create potential computer errors connected to the storage of calendar data for dates after the year 2000, making the year 2000 impossible to tell apart from 1900.)

In the early twenty-first century (the world didn't go back a century), I was with my son Jack, my brother John, and friends who were working on a space program years before retiring. The talks were about the space program, which I also was involved with once. I was the lead engineer on a special Apollo-Soyuz Space Module VHF/FM antenna array. (By the way, the Apollo-Soyuz Test Project was the first international space mission.) This was the basis for thoughts about all things having to do with space and our "Solar System" (the gravitationally bound system of the Sun and the objects that orbit it.)

We often sat outside in the evening, talking about all kinds of things. One year, I brought my iPad (a tablet or slate computer) with a program that showed the planets (there are eight in the Solar System, each circulating the Sun at a different distance) and stars based on the position of the pad. Imagine the Sun as a clock and the planets as its hands circling it—it's just a metaphor—as an image of our Solar System.

And the iPad was a good tool to pan around to find which direction the planets were in at that time. My nephew Paul was looking down through the earth with it and said, while behind me, "I see Uranus." This was the type of fun we had and I continue to have with my edgy ideas about the Universe and its existence or extinction.

Down-to-Earth Topics to Talk About, Seriously

So I digress. Now, Jack and I had two key subjects to discuss. Under the bright light of the Full Moon with a skyscape of stars, it went like this: Anything we came up with to explain things like gravity and Dark Energy needed to be logical. This brought us to the features of gravity we knew of without second thought. It is invisible and can affect mass. It can pass through any mass we know of. And finally, the effects we observe between two masses follow the inverse square law of gravity. That is, if the force we measure between two masses that are a quarter inch apart is 1/4g, then when the distance is doubled to half an inch, the force will be 1/9g. The reverse is also true. If we measure a one-inch-distance force between two masses and then move the masses to half an inch apart, the force we measure is four times the one-inch measurement. And that's not all . . .

According to some science pundits, "Being strictly geometric in its origin, the inverse square law applies to diverse phenomena. Point sources of gravitational force, electric field, light, sound or radiation obey the inverse square law." (4)

Simply put, the statement "Being . . . geometric in its origin" means it's based in the physical laws of a sphere's surface area as it relates to its radius. That means it is derived from the mechanical properties of a sphere's area versus its radius. The point? It is a very basic and simple mechanical phenomenon, which is key to understanding the concepts and logic in this book.

So gravity follows the inverse square law. The interesting thing is that if two magnets' forces are measured in the same manner, they also follow the inverse square law. This brought us to the exciting conclusion that we need to find out what magnetism really is because it is so much like gravity. They both seem to pass through mass easily. They both have the same force characteristics on like mass, and they are both invisible. Magnetism is the key focus of this book because in the beginning of our search for answers, it was the least understood. As it turned out, this was the correct approach.

As of now, the search for Dark Energy has ended in the discovery that magnetism can explain what we see as the effects of Dark Energy. In other words, this means the energy associated with Dark Energy, "a strange form of matter that only interacts with 'normal' matter through gravity," is the magnetic effects of the Dark Matter on its surroundings.

Defining Dimensions

The topic of dimensions is a broad one. It can be explained to a child, teenager, or adult. In plain English, space is measured in three dimensions: A dimension in physics is a measurement of length in one direction. Samples: Width, depth, and height are dimensions. One-dimensional is a line. Two-dimensional is a box. Three-dimensional is a cube. So there you have it—a simplified definition of dimensions. And there's more . . .

Controversy exists about whether we live in a one-, two-, or three-dimensional Universe. Back during Albert Einstein's work on physics, time and space were viewed as different dimensions. Einstein's theories have shown that due to relativity of motion, space and time can be mathematically combined into space-time.

Into the World of Cosmos Dimensions

On the subject of being able to pass through any mass, we had a lot of discussion and thinking with no evidence to support what we were coming up with. Many idea dead ends have happened through this effort. There ended up only one thought that seemed to fit both the invisibility and the ability to pass through any mass: the idea that mass could be one-dimensional.

Now, I am thinking about one-dimensional mass and what it would be like. My thinking went like this. I know of three-dimensional mass but not two- or one-dimensional mass. Logic says that if there is one-dimensional mass, there must also be two-dimensional mass. Also, logic tells me that any one-dimensional mass must be by definition very straight, like a laser beam, but there is no width or height to measure. This laser beam is at its basic size smaller than the parts of an atom, and it will slip between the electrons (packets of electromagnetic energy) and the nucleus without significantly disturbing it or the atom. This would only apply to when it is moved through mass along its length, like a metal rod I push into sand straight down, only I could not actually hold on to it, as it would pass through me as well as the sand with ease. But for now, in my mind, I am able to hold it and push it. There is so much room between the electrons and the atom's nucleus that it could be slightly curved and still have the same properties. But there's more . . .

If I had a small—close-to-zero-but-greater-than small, aka "infinitesimal"—diameter rod in the sand and tried to move it to the left or right or forward or backward, I would feel resistance. This notion fits with a phenomenon that I am aware of. Here's proof:

Imagine seeing your neighbor's electric automobile using solar energy. Well, it uses eddy-current brakes, right? There is an aluminum disk that spins with the wheels and has an electromagnet, or magnets, very nearby the disk that is fixed to the car frame. Without the electromagnets energized, the car rolls easily. When the car is moving faster than a few miles an hour and the electromagnets are energized, the car slows because of the friction between the aluminum disk and the invisible magnetic lines of force now passing through the aluminum disk. Also, the disk heats up due to this friction with the invisible magnetic lines of force. This is akin to the one-dimensional rod in the sand and me trying to move it left and right while feeling resistance. And that's not all.

I've demonstrated this to myself in my garage. I have a drill-motor-mountable 12-inch-diameter aluminum disk, a drill motor, and a neodymium magnet. When the disk is not spinning, I can move the magnet all over the disk, touching it with ease. If I set the disk spinning, I cannot touch the disk with the magnet. It will not let me do it. So where is this analogy I'm making taking you, other than on a joy ride?

This phenomenon is why I believe that the magnetic lines of a permanent magnet or an electromagnet are one-dimensional in their basic nature but curve into the second and third dimensions to compound their individual forces into a more compact, yet greater, flux/force, which presents to the 3-D observer as the commonly understood characteristics of mass. And by extension, all magnetic lines are one-dimensional in their basic structure.

The Genesis of Shaylik

It's time I now introduce to you the use of the concocted word "Shaylik" to describe not only one-dimensional magnetic lines but two- and three-dimensional magnetic lines. This invented word is named after my granddaughter, Shayla, who agreed I could do this and share her given name, which I gave a cosmos spin since she is part of my Universe. Here is my definition of Shaylik and why it is important to me and should be to you too.

The definition of Shaylik (my word) isn't in the dictionary. However, that doesn't mean it's nonexistent. Shaylik describes the fundamental building blocks of creation. It is the central unit making up all there is or will be in our Universe; and through eddy-current effects, it's the reason for gravity.

Scientific evidence shows Shaylik as flux rope (a series of loops found on the sun), flux lines (magnetic lines of force starting from the North Pole and ending at the South Pole), and flux tubes (isolated concentrations of magnetic field found in the solar convection zone). And yes, all of these are similar to Shaylik but are related to specific environments, such as the Sun (Earth's star) or between planets or where current flows in a wire.

Shaylik is a much broader term that includes all environments and circumstances and is the basis for these other concepts and observations of magnetism. The word "Shaylik," because it can be one-dimensional, also explains the features of magnetic lines of force, such as invisibility, mass attributes, and environmental effects.

So a simple thought—gravity is related to magnetism—began our mental journey. We understood that whatever we developed from this point on would not be easy to find. If it were easy-peasy, then our scientists would know what Dark Matter and gravity are—and this book would just be old news.

We also knew that this search for the truth would take a long time, possibly beyond my lifetime, and would relate to all that there is in our science and math, and our Universe. This information would be a search for something that is fundamental to all that exists in our Universe and has escaped our understanding for all of our history.

Finally, we did this father-son deep discussion under the bright night sky because, for us, two science buffs, this was fun and would sharpen our minds if nothing else. As it turned out, it has been fun and exciting at times. Many nights of lost sleep and many pleasant surprises keep me going because I am constantly pondering the missing puzzle pieces, big and small, of the cosmos.

Welcome to Cosmology

There is a branch of science that best describes what we are interested in. It's called "Cosmology." No, it has nothing to do with what people put on their faces. Here's one description. It's the branch of philosophy dealing with the origin and general structure of the Universe, with its parts, elements, and laws, and especially with such of its characteristics as space, time, causality, and freedom. It is also the branch of astronomy that deals with the general structure and evolution of the Universe.

So this is a huge subject covering everything from the smallest phenomena we know of, the parts of the atom, to the largest phenomenon we know of, the structure of the Universe, and everything in between.

OK, here's the real deal in nonscientific-ese. No one book can cover all that is involved in our Universe. However, this book will focus on one area that is important to the understanding of most phenomena in our Universe. I will also do my best to keep it easy to understand, but there will be times when I cannot figure out how to simplify a concept.

What's more, this book is intended to appeal to all people of all ages who are interested. Those from the "just curious" to the scientist should be able to understand my scientific-ese if I have met my goal. If not, my apologies to those of you I may lose along the way. So whether this book is on the mark or only partially correct and truthful, my purpose is to enlighten you along this journey of the cosmos. Enjoy the work and look forward to other forward thinkers to pave the way to work that remains to be done. I am sure many other surprises are out there to discover.

Meanwhile, take a look at chapter 2, "Chiral-Shaylik Connection," to see how my theories link and are like the Legos toy (kids use these blocks to build things), which are essential to putting the bits and pieces of the big evolutionary puzzle together to see our Universe in bright light.

NEW THOUGHTS, OLD UNIVERSE CLUES

- ✓ Without a doubt, gravity is an open book without logical explanations—and it needs to be scrutinized more.
- ✓ Dark Matter and Dark Energy are essential forms of energy and matter that do matter in the Universe.
- ✓ Magnetic lines are multidimensional and serve a purpose to the Universe.
- ✓ "Shaylik" is a new word that has real meaning in the building blocks of the Creation.
- ✓ Cosmology ideas with the structure of our Universe matter more than you may think they do.

CHAPTER 2

Chiral-Shaylik Connection

It is reasonable that forces directed toward bodies depend on the nature and the quantity of matter of such bodies, as happens in the case of magnetic bodies.
—Isaac Newton

As time passed, I didn't stop with my fascination of Shaylik—but I noticed a real name (it has an indirect link to my magic keyword) in two real branches of science that commonly use the term "chiral" or "chirality." They are the sciences of Biology and Solar science. It's interesting, and telling to me, that this word is used in such different disciplines. Biology is about life on our planet. Solar science is about plasmas and magnetics. So the million-dollar question is, could there be a connection between the two sciences and Shaylik that is not currently recognized? Time will tell. Some research has been done on chirality, and my observations confirm what I have read—and I add my own touch to findings.

Definition of Chirality

Chirality/kaɪˈrælɪtiː/ is a property of asymmetry important in several branches of science. The word "chirality" is derived from the Greek χειρ ("kheir"), "hand," a familiar chiral object.

Chirality Under the Microscope

Did you know "chirality" is an important word in the human medical field? The thing is, it made its way into being significant when used as the antinausea drug thalidomide. It was introduced for pregnant women around the 1960s. The medication was tested and approved for use without the knowledge that it had a required chiral nature. One chiral version of the drug, in fact, was given support with no consideration of the other chirality. The approved version was then produced by many manufacturers without consideration of what chirality it had. And then there were consequences.

Only two chirality results are possible. It was thought that there would be no difference in its effects. So some manufacturers made left-handed thalidomide and some made right-handed. Unfortunately, one of them caused serious deformation of the fetus in pregnant women who took it for nausea. As a result, it was removed from the market for further testing, which showed that the chirality of the test version was the opposite of the chirality of the version that caused fetal malformation—a human tragedy at the time and one that is not forgotten. So without a doubt, chirality can be important in human chemistry. But that's not all . . .

Chirality is also a major component in the way things work together, or not, in the Universe. When I think of chirality, I envision a corkscrew. Picture the handy gadget you use for pulling corks from bottles. It contains a spiral metal rod that is inserted into the cork, and a handle that extracts it, right? Well, all of them are made right-handed because most humans are right-handed and it is easier to use for a right-handed person. Compression springs that we can buy at the hardware store are also typically one chirality, at least by batch, because they are made with machines that are identical. An interesting feature of the compression springs is, they become entangled easily. If you need a couple of springs from the bin at the hardware store, just taking one out of the bin is almost impossible. They are all tangled together, so I have to untangle them to get just a couple of them. Shaylik is like this. This is a type of entanglement that is strictly mechanical.

Shaylik Comes in Two Chiralities

Interestingly, Shaylik comes in two chiralities (plural form for "chirality") and only two. If the chiralities are the same as well as the frequency, then the Shaylik will mechanically entangle and stick together mechanically, like the compression springs. In this case, the chirality is the same, left or right, as well as the turns per inch, which is akin to frequency. So under the right conditions, Shaylik can entangle with other Shaylik and stick together. Conversely, when Shaylik is of different chiralities and/or different frequencies, they will repel each other for mechanical reasons. Most of the time, Shaylik cannot combine and is repulsive to other unlike Shaylik. This is because both chirality and frequency correlation are required, which is the least-likely condition.

The DNA double-helix (or spiral) of our DNA (or the hereditary material in humans and organisms) is chiral. In fact, all life on Earth has chiral DNA in its cells. Most of the DNA on our planet is homochiral and left-handed. There are exceptions. For instance, bee honey is right-hand chiral, which may be why we do not digest honey like we do other foods. The molecules of our planet can be either chiral or not. I suspect this is true all the way down to the components of the atom.

It's known that electrons can be different and can briefly exist in our environment as positrons. Only two possibilities exist. One is charged negatively, and the other is charged positively. This only-two-state possibility is similar to chirality. I have wondered, "Can they be the same thing but with different definitions?" Or, "Is it imaginable Earth, or the whole Solar System, may possibly have an environment that only supports one magnetic homochirality and suppresses the other?" Scientific "food" for thought is my answer because, in the cosmos, anything is conceivable until proven otherwise. It's arrogant to think the Universe and its wonders have been solved when it is like a puzzle that needs to be completed. No, it's still incomplete like a jigsaw puzzle with missing parts.

Untwisting Shaylik and Chirality

Going back over to Shaylik and chirality, I haven't discovered a single instance where a description of some magnetic phenomenon is not abnormal or twisted. To be precise, the strange effects point to chiral. It's never just straight lines of force; it is always twisted—twisted single lines and twisted double lines, but always twisted. What's more, this is true at all scales I have investigated, from structures larger than galaxies to the atom (molecule or tiny part) and smaller. So when I say magnetic lines of force, I assume it means twisted magnetic lines of force. I can find no other explanation that makes sense. So all magnetic lines are chiral and must be either right- or left-handed. It's just the nature of the beast.

The Magic of Chirality: Magnetism

Since I don't know of anything in our Universe that is not magnetic, I see that all there is will basically come from and have features that relate to chirality and magnetism. Sure, some molecules are not chiral in their nature, but the molecules' constituent atoms and related magnetics are chiral.

Until now, I was thinking magnetism was a phenomenon separate from things made of mass. Previous conventional thinking was that mass is fundamentally different from magnetism. After all, magnetism is invisible and can pass through mass easily. Mass is visible and can't pass through other mass except in a few instances, like neutrinos (particles). Now I see that mass is simply bundled magnetism, bundled like the DNA of a living cell, all twisted and knotted together to make the electrons, protons (atomic particle found in the nucleus like a neutron but has a positive charge), and neutrons that are mass.

So now I imagine a pipe-like structure that is magnetic and is made of two ropes that are twisted themselves and twisting around each other. There is a distance between the ropes that is constant because the two ropes repel each other and the entire structure is under tension, which holds them in at a constant distance. The forces inward are exactly balancing the forces outward.

Because of this structure, it would be more like a ribbed pipe than a smooth one, and there are lower-pressure magnetic gaps between the ropes, so it would be a leaky spiral pipe. If this pipe were spinning fast enough, it could trap three-dimensional mass inside it. Mass like the plasma on our Sun has some inertia characteristics and will not be moved easily. These pipes could be miles wide inside. So this trapped plasma, which is essentially ionic matter, would be forced to the middle of the pipe and would be visible to our telescopes. This means we could see the shepherded plasma because the pipe that is holding it inside is invisible and transparent.

Besides, it follows that a ribbed spinning pipe that has plasma inside would act like a pump, moving the plasma in the same direction the spirals are moving. This, then, would move the plasma from one location to another as long as there is a plasma supply at the input to this pump and a low-magnetic-pressure opening at the other end of the pipe. Wow, this double-helix Shaylik is very versatile with a capital "V." It not only can be a rope or a pipe but can also be a plasma pump. In fact, the Shaylik can be anything the environment will allow to be stable.

On our Sun, these visible lines that are surrounded by invisible Shaylik, helical spinning pipes are what our scientists call a group of solar-related words, including "barbs," "coronal loops," "coronal ropes," "fibrils," "filaments," "prominences," "plumes," and "sigmoids." In other words, this form of Shaylik explains all the linear structure we see on the Sun. From Shaylik to the Sun is intriguing. Look up into the sky and, on a clear day, you will see the big yellow ball in the sky and, now you know, Shaylik is busy at work even if you can't see it.

10 Puzzle Parts to Legos and Shaylik

What if there was a very special set of Legos linked to Shaylik? It's not the set you had as a kid but something new, improved, twenty-first century-ish. This set of Legos is made of only two parts, and that's it. Read on and discover the link between the building blocks and Shaylik.

1. These two Legos look like a long, narrow spiral spring. The only difference between them is, one is twisted to the right and the other is twisted to the left. There is also an unlimited supply of these parts. You could have 100 lefts and 100 rights at a time. And you could keep going back to the box to get 100 of each, every time you needed to. These Legos have some special characteristics.
2. For starters, the right Legos are red and the left are blue. The red Legos will stick to each other like magnets, and the blue ones will also stick together like magnets. But when you or I try to put a blue Lego together with a red one, they push back on each other. They do not attract one another like a mismatched dog and cat or man and woman.
3. Next, I can stack the blue group into a long string of them, and I can do the same with the red ones. I will call these two assemblies the "red string" and the "blue string." So now the red string is tough to break apart with my hands, but it can be done. If I lay another red string of these Legos next to the first one, why, they will come together easily all by themselves! The same thing will happen with two blue strings of these Legos. So for both the red and blue groups, I can make an assembly of two or more of the red or blue strings. As I do this, the Legos become harder and harder to pull apart. And that's not all . . .
4. Then I build an assembly of 100 red strings and 100 blue strings that are about 12 inches long. I can add a red 12-inch string to a second one and make a string 24 inches long. I do this with both the red and the blue Legos, still only using two basic pieces. So now I have 50 of each that

are 24 inches long, and they are all lined up in parallel to each other and all the red ones are stuck together, as are all the blue ones.

5. Now I have the beginnings of a rope in each color. All I have to do is twist them together and I will have two ropes, one red and one blue. When I try to twist them together, I find that the red ropes will only allow me to twist them in the same direction as the original parts. This is the only way the red or blue ropes will allow me to twist them together.
6. So the twisted red and blue ropes are all collected and twisted such that all of the parts have the same twist. All the portions that made the red rope are twisted to the right, and all the parts that made the blue rope are twisted to the left. Now I have two ropes that, because the strings are all twisted together, I cannot pull apart because I am not strong enough. All these parts working together are now so strong that I cannot separate them.
7. Next, I realize I could have made the ropes fatter and shorter if I had wanted to, and/or I could have made them longer and narrower. Because I have an unlimited supply of these two types of Legos, there is no limit to how long or wide I wish to make them.
8. Another thought is, what happens when I twist the red rope with the blue rope? I know that what the red Legos and the red ropes are made of did not like to be near the blue Legos or what the blue ropes are made of. They are not compatible. So I try to do this experiment to see what happens. I tie the red and blue ropes I have to the same post and begin twisting them together. I find I can twist them together with either a right or a left twist, and no matter how much tension I use to pull them together, they will not touch each other. There is always a gap between them. The tighter I make them twist together, the closer the red and blue ropes are, but I cannot get them to touch each other.
9. I twisted the red and blue ropes, so they stay in a straight line, like a pipe. The end that is not tied anywhere is just floating there in front of me in my hand. I mull over this a bit and realize this assembly is acting like a pipe—not a rope. In result, these are the forces between the tension in the ropes and the repulsion between them, coupled with the fact they are twisted around each other. In other words, it has distributed the forces evenly along the length of the red-and-blue assembly such that it is acting like a pipe instead of a rope. So now, from just two Lego pieces, I have made strings, which then made ropes, which then made a pipe-like structure that is very straight. I have something that can be flexible like a rope or rigid like a pipe depending on how it is put together. Fun stuff! But there's more.
10. What I have not told you until now is, the Legos are not man-made. In fact, they are so small that we cannot see them with even our first-class microscopes. These two Legos are not made by anyone; they just are there to become whatever they want to try to be. I think of it as if there is an unlimited amount of them and they are in a box that is shaking. They are just being jostled around in the box and can come together in any way they want to. Whatever they come up with will be a stable structure in the current environment or unstable and not survive.

In brief, this is what I call Shaylik. It's basically two-dimensional and is what's called magnetic lines of force—and it's all chiral.

The double rope is representative of the double-helix configuration that exists between all matter. It can exist in all the environments of our Universe because it is the basis for all that there is. One of the results of this is what we call water, which, interestingly enough, can be either a fluid or a rigid structure depending on the environment it is in, just like the basic structures it is made of. Another outcome of this is the DNA double-helix found in the nucleus of all living cells.

Next up is a different topic, but I'm bringing along Shaylik. We're going on an adventurous journey up high in the sky to mighty Mars. You'll discover what type of storm eruptions happen on this planet—and why it matters.

NEW THOUGHTS, OLD UNIVERSE CLUES

- ✓ Chirality is a key and critical feature of magnetic phenomena—good and bad.
- ✓ It is a real word and goes back to the twentieth century when it is linked to a harmful man-made drug . . .
- ✓ But chirality is also a major component in the way other things work in the Universe.
- ✓ Analyzing Shaylik and chirality takes scrutiny to untwist its complicated nature.
- ✓ There are only two possible configurations: right-handed and left-handed.
- ✓ Shaylik is the universal building block for all that there is and all that there could be in our Universe.
- ✓ And note, chirality and magnetism are bundled like the DNA of each living cell.
- ✓ The building blocks of Shaylik are less complicated than you think if you consider Legos and the interesting link.
- ✓ Both chirality and Shaylik are key players in nature.

PART 2

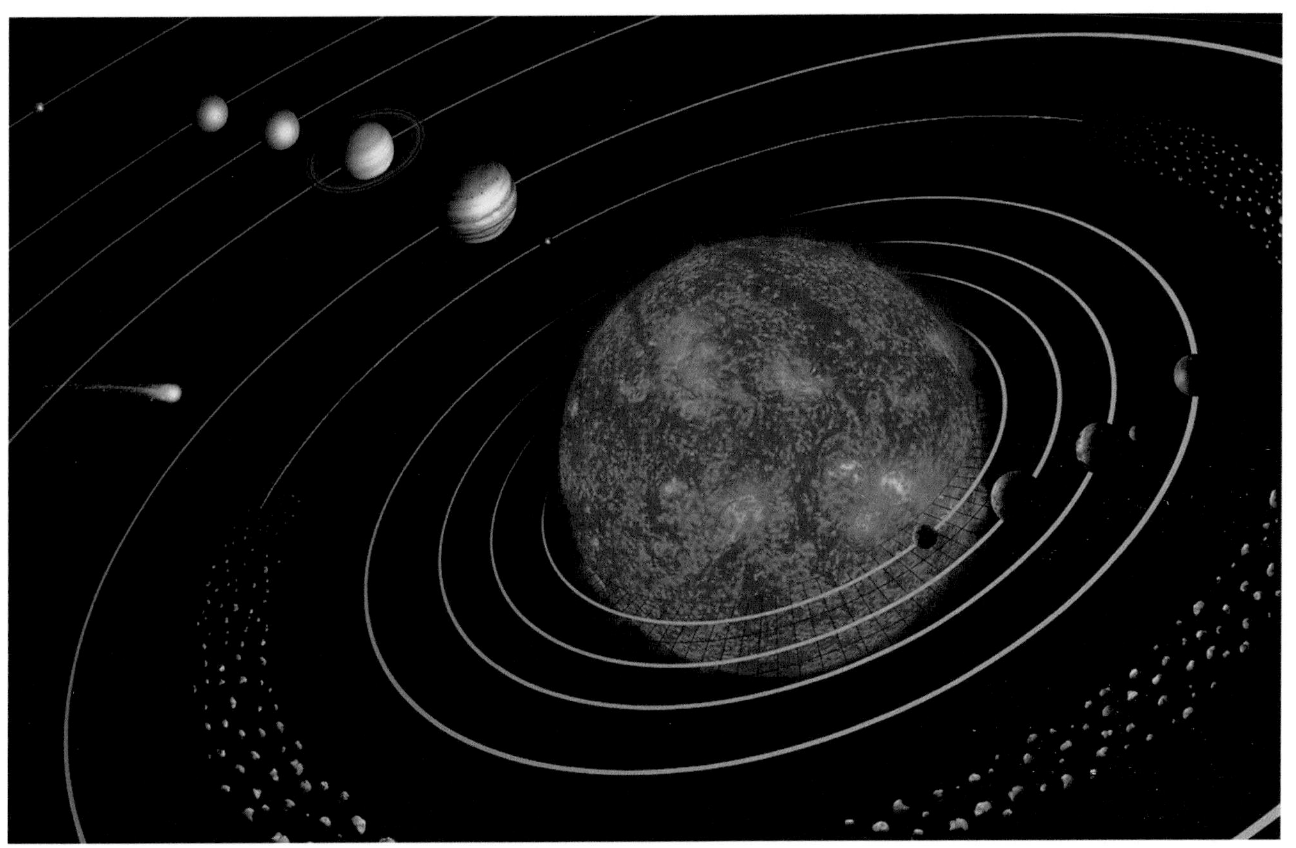

SHAYLIK AND PLANETS

CHAPTER 3

Dust Storms on Mars

The great earthquake shall be in the month of May;
Saturn, Capricorn, Jupiter, Mercury in
Taurus; Venus, also Cancer, Mars in Zero.
—Nostradamus

Storms in the cosmos is an interesting happening and one that has been on my brain from time to time. My son Jack and I have noticed that what we are mulling over with fascinating "should" and "could" questions—including cause of weather on Earth or planets—can have plausible answers. Our ideas, both usual and unusual, don't stop anywhere in the Universe. In fact, our science-fiction-type musings may even be the seeds for weather on a planet with an atmosphere.

Taken By Storm (or Mars)

Tornadoes and hurricanes are chiral, just like Shaylik. There was a time, at our location in Arizona, when we had unusual periods of active all-night thunderstorms for many days. Now, while this strange phenomenon was occurring on Earth, Earth was near conjunction with an outer planet (Saturn) that was in conjunction with a nearby star (HIP 102780). The effects of a union, like this, would be greatest at nighttime when that side of Earth was facing the planet and star in question. That time of year was when the summer monsoons happen, and they almost always happen in the afternoon or later, stopping near sunset. But on this one unforgettable occasion, something was different. And I'll tell you why this happening was so out of the ordinary.

It's well known that Mars and its global dust storms (or sandstorm, a meteorological phenomenon often happening in desert regions) happen when Mars is closest to the Sun and during the springs and summers of the planet's southern hemisphere. The southern hemisphere also has the most flat land. These factors are clearly contributors to the phenomenon. There are dust storms on Mars all year round, but the global storms are obviously seasonal.

Mars has a very thin atmosphere over its surface. Its surface is like talcum powder, and it is mostly iron. As most of us have experienced, iron loves magnets. So it seems logical that any turbulent magnetic forces would easily pick up this iron talc and cause friction in the atmosphere. Friction in the atmosphere could cause static electricity, which would help lift the iron talc off the surface and into the Mars atmosphere. Mars is a unique environment, which may be significantly more sensitive to magnetic turbulence. And that means the planet is more vulnerable to severe weather.

So the predictions of Mars global dust storms are based on the seasons, which is very unspecific date-wise. There is another factor Shaylik can add to the idea and is more date specific. That is simply pinpointing the Mars conjunctions with the Shaylik from nearby stars. This is when Mars is between the nearby stars and our Sun. So Mars is passing through the double-helical Shaylik from star to star.

A perfect conjunction would be when they are exactly in line. I have analyzed the nearby stars and the Mars orbit using the AstroGrav software program. (It is a Solar System simulator that analyzes the gravitational interactions between astronomical bodies, so that the motions of asteroids and comets are simulated more accurately than with planetarium applications.) I (and you too) can set this computer program to look at where the planets and stars are from the perspective of our Sun. So it's as if I were standing on the Sun while looking at Mars and the stars through my telescope.

Therefore, with the entertaining software set up for view from the Sun, I used a criterion of 0.25 degrees from perfect alignment of Star/Mars/Sun. I also tried to focus on named stars, as they are best known and close to Earth. The following graphic shows the conjunctions that are within 0.25 degrees of perfection by their distance and Mars date. The Mars date system used by some astronomers is different from Earth's. The designation for a Mars date is "Ls," and there are 360 days in a Mars year. Ls 1 was on the Earth day Feb 10, 2021, for one example.

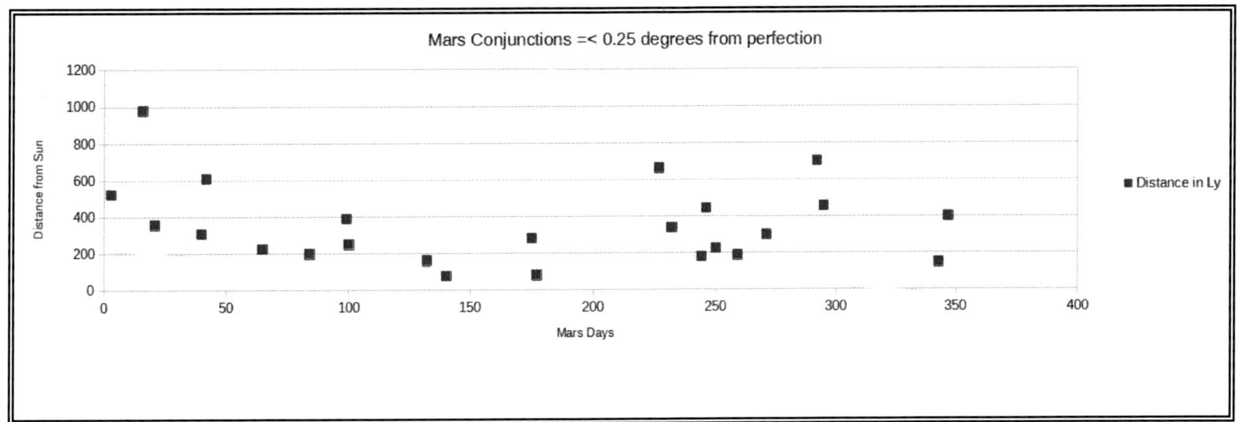

Author generated graphic

Imagine Mars year by day along the bottom, and the left side is in light-years of distance. Each blue square is a star from the list that follows. So the bottom left corner is day 0 and distance 0.

From Ls 227 to Ls 295, there is a conjunction every eight days, on average, over 68 days. From Ls 340 to Ls 180 is 200 days in which there are 15 conjunctions, or one every 13 days on average. Mars global dust storms tend to happen around Ls 250, between Ls 225 and Ls 310.

These star conjunctions happen every Mars year and provide an exact Mars and Earth date to look for Mars global dust storms as well as nonglobal storms. Future Mars phenomena will show if these conjunctions are a good predictor or not. If this is the case, then the concept of Shaylik between stars and what happens when a planet passes through that Shaylik will be on its way to acceptance. So far, this seems to be historically correct.

This would infer, by logic, that Shaylik can affect the weather of at least one planet and possibly more. It also could verify how straight the Shaylik is between stars.

DATES FOR DUST STORMS ON MARS

The following is a list of Mars dates to look for dust storms and which star conjunction is involved.

Mars Day (Ls#)	Earth Date	Earth Time	Distance in Light-Years	Star Name
02	Dec 31,22	1100	522	132 Tau
16	Feb 04,23	0100	980	ω Gem
21	Feb 08,23	2100	357	48 Gem
40	Mar 22,23	0500	308	η Cnc
42	Mar 26,23	0500	610	39 Cnc
65	May 17,23	0001	225	34 Leo
84	Jun 28,23	1500	198	σ Leo
99	Aug 03,23	1300	389	13 Vir
100	Aug 04,23	0100	250	Zaniah
132	Oct 12,23	0835	163	Kambalia
140	Oct 28,23	2250	77	Zubenelgenubi
175	Jan 4,24	0900	281	o Oph
177	Feb 19, 22	1100	81	44 Oph
227	May 14,22	0100	663	20 Cap
232	May 21,22	1300	338	31 Cap
244	Jun 09,22	1700	180	ι Aqr
246	Jun 14,22	0100	445	42 Aqr
250	Jun 20,22	1200	225	58 Aqr
259	Jul 04,22	1100	188	83 Aqr
271	Jul 23,22	0100	299	20 Psc

292	**Aug 25,22**	**2100**	**702**	**73 Psc**
295	**Aug 30,22**	**1100**	**456**	**88 Psc**
343	**Nov 23,22**	**1600**	**147**	**k1&k2 tau**
347	**Dec 01,22**	**0500**	**401**	**t tau**

There's a connection of the seasons of Mars global dust storms and the stars Mars is in conjunction with over a Mars year. So it's likely that this phenomenon is a contributor to the Mars weather—and by extension, other planets' weather. By the way, Mars's orbit will repeat this sequence of star conjunctions every year for at least the next 10 Mars years.

If I could discover the dates when Mars dust storms are first seen, I would compare them to the dates above. If the dates are always on, or one or two days after these dates, then there would be a correlation that supports this concept. If the dates are random, then this would be evidence I'm not correct in this thinking. Only time will tell!

Sounds of a Shooting Star

Before the Sabrina fishing years, the Westberg family packed up the car and went on a road trip into the Mojave, California, desert for a weekend of motorcycle and dune-buggy fun. We were made up of four families with all ages of youngsters.

One night, a small group of us strolled out into the wasteland to stargaze. The area was sparse. No houses. It was used by other campers over the years. There were above-ground power lines nearby, and trash remnants were left from previous campers in the area. Chewing gum wrappers, tinfoil, paper, cans, and bottles were spread around us.

While looking at the stars and debating what they were named, we saw a shooting star come right over us. It couldn't have been more than 10 miles above us and probably about five miles up. It wasn't exceptional in size and gave us five seconds of "ahs" and "ohs" and the wonder.

Something unusual happened at the same time we saw the shooting star. We heard sounds with the timing of the shooting star's visual passage. The area we were in was quiet, and we all heard the same thing. A rustling sound surprised us. We all agreed it happened right when the shooting star passed over us. How could this be?

If it were a noise from the shooting star, it would have happened after the star passed, but it happened *while* it was over us. Lots of discussion ensued, and we came to the conclusion that the sound came from the shooting star and it must have been a magnetic phenomenon, which could happen simultaneously.

We thought a magnetic wave came over us and affected the metallic trash debris all around us, which made the rustling sound.

Since that night, I've learned that a shooting star is a rock or stone called a "meteor" or "meteorite" in space headed for Earth. I sense I'm hardly alone, and perhaps you, too, have experienced the magnetic-wave noise incident. Ah, the wonders of sound and space.

P.S. If anyone reading this book has had a similar experience, I would like to hear about it.

If you think dust storms on Mars are exciting, it doesn't stop there when thinking outside-in. What about ice volcanoes? Read on—and in chapter 4, some new notions may enthrall your mind too!

NEW THOUGHTS, OLD UNIVERSE CLUES

- ✓ Storms are chiral, much like Shaylik.
- ✓ Global dust storms occur when Mars is closest to the Sun. AstroGrav can help to look at the planets and stars.
- ✓ There is a correlation and concentration of nearby stars during the known period of Mars global dust storms . . .
- ✓ Logic conveys this idea needs further observation for particulars as far as dates of origination of Mars dust storms. Tracking dates for dust storms on Mars may lead to better weather forecasting on Earth.
- ✓ If this theory is shown to be correct, then we must also look to Earth for similar phenomena.

CHAPTER 4

Ice Volcanoes at Ceres

In all chaos there is a cosmos, in all disorder a secret order.
—Carl Jung

By the time my son Jack and I had formed the basic concept of Shaylik—and its link to dust storms on Mars—we were on to a different cosmos adventure. Let me introduce you to ice volcanoes, also called "cryovolcanoes," which are on the minor planet Ceres.

We both are constantly looking for confirmation of our theory about the effects of Shaylik. As is common, we first looked on the Web for a reasonable explanation for this phenomenon, and we already had our own ideas. Not only did we not find a reasonable explanation, but we could find no explanation. This is not to say there is not one out there somewhere, but we could not find anything that made sense.

No, cryovolcanoes don't spew hot magma like volcanoes on Earth. Forget images seen in the film *Dante's Peak* or *2012* (no mountain-town meltdown or zapping the Hawaiian Islands). However, other planets and moons in the Solar System do have cryovolcanoes. And these volcanoes erupt ice liquid such as ammonia and water.

An ice volcano on Ceres, however, would have no smell or sound because there is no atmosphere. It would initially be like a slushy from a convenience store and then become hard frozen. Earth-based cryovolcanoes would have an oozing sound.

Ceres, a Minor Planet with Major Stuff

A cryovolcano can be found on the minor planet Ceres, which is in orbit around our Sun between Mars and Jupiter. There are photos to show you where the planets and some minor planets are located relative to each other. Often, a poster will be on the wall of a high school or college astronomy class. These days, there are programs to show you past and future dates. (1)

Minor planet Ceres caught my interest when I read that there were the rare cryovolcanoes apparent on the surface of this planet. The article also said that this is related to salt water inside the planet. A cryovolcano is simply one that happens where it is cold like the space is around Ceres.

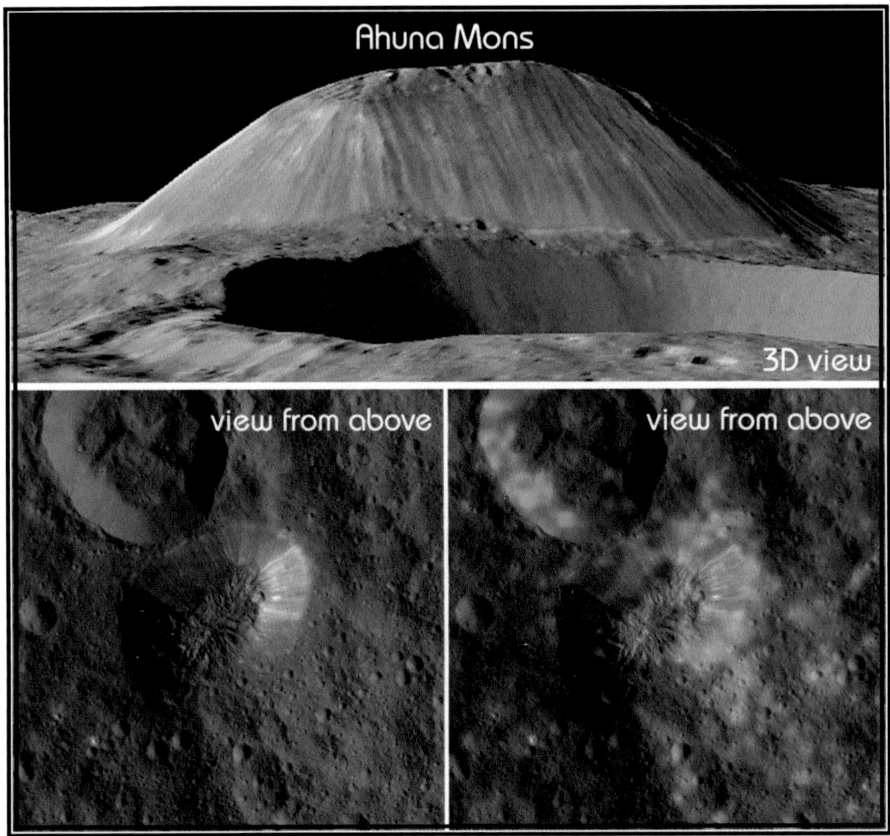

Nasa Pictures Ahuna Mons - Courtesy NASA JPL

Ahuna Mons: Big Mountain, Big Activity

Did you know Ahuna Mons is the biggest mountain on the small planet (and asteroid) Ceres? Actually, it is the one and only mountain on Ceres. Brilliant colored lights shroud its slopes, and these streaks are believed to be salt, like Cererian bright spots. This, in turn, happened from ice volcanic eruptions straight from Ceres's crater.

OK, wait a minute. This minor planet is located far away from any other planet in the cold of space. There is no apparent gravitational reason for the planet to heat up. I can understand why moons that are near their planets could be heated by the stresses of gravity from their nearby planets and other moons. But guess what? No such things are true for Ceres. Temperatures in the asteroid belt are from -73 degrees C to -108 degrees C. So how in the world is this possible?

The dwarf planet Ceres orbits our Sun in about 4.6 Earth years. Ceres's orbital inclination is about 10.6 degrees. Earth's is about 7.15 degrees from the ecliptic. So the stars that Ceres could come into conjunction with are much more numerous than those that Earth encounters year to year.

I don't know which star Ceres could be in conjunction with or which one is the largest and closest. But I can speculate based on the stars Earth encounters. The first star that came to mind is Lambda Aquarii. Next was Alpha 1 and 2 Librae. Lambda Aquarii is 3.6 solar masses and 365ly away, while Alpha Librae is actually two stars totaling about 3.4 solar masses, and they are only 76ly away. Earth comes into conjunction with both of these objects. The winner in my mind is Alpha Librae because the two stars are about the same mass as Lambda Aquarii alone, and they are much closer.

Another factor is, Ceres orbits the Sun 4.6 times slower than Earth. So when there is a conjunction, it would last 4.6 times longer. This is the kind of thinking that will yield the possibilities for Ceres conjunctions that are significant and could explain the cryovolcanoes. Ceres passing laterally through double-helix Shaylik that is flowing to our sun would offer some resistance and cause temporary heating of the dwarf planet. The effect of magnetic lines heating mass it is passing through is known as "eddy currents."

This is more evidence of the Shaylik that exists between stars and its heating (eddy current) effects. There is no other explanation I can think of for this occasional cryovolcanic phenomenon on Ceres.

On the minor planet Ceres, occasional cryovolcanoes are caused by its passing through magnetic lines of force from stars or planets or both. Check out online programs to find out where the planets and some minor planets are located relative to each other and past and future dates.

I've read one explanation of the reason for occasional Ceres cryovolcanoes as the heat coming from the radiation of radioactive elements inside the planet. It is possible that there is this type of radioactive heating going on inside Ceres. The problem I see with this is, as far as I know, the heating from this type of force cannot occasionally happen. Once the planet was established, there would be a fixed amount of radioactive matter, which will only degrade over time and not have spurts of heating energy. The atomic radiation would only get less and less over time and not have higher energy output over time on occasion.

While we are on the topic of Earth changes and planets, it's time to revisit gravity. And this time around, let's take a peek at new visions of forces on the horizon. Instead of rolling your eyes, you may discover mind-boggling musings about gravity.

NEW THOUGHTS, OLD UNIVERSE CLUES

- ✓ There is no definite rhyme or reason for why ice volcanoes happen on Ceres . . . but if you look at possible reasons, it's intriguing.
- ✓ Rare cryovolcanoes are a rare phenomenon and may be linked to the placement and timing of stars.
- ✓ Heat from the Sun and magnetic heating also may trigger an ice volcano.
- ✓ Also, these beautiful volcanoes may happen because of magnetic lines of force from stars and planets.
- ✓ Further research is needed to discover all the triggers of ice volcanoes. For now, these reasons are a start to pinpoint the source.

PART 3

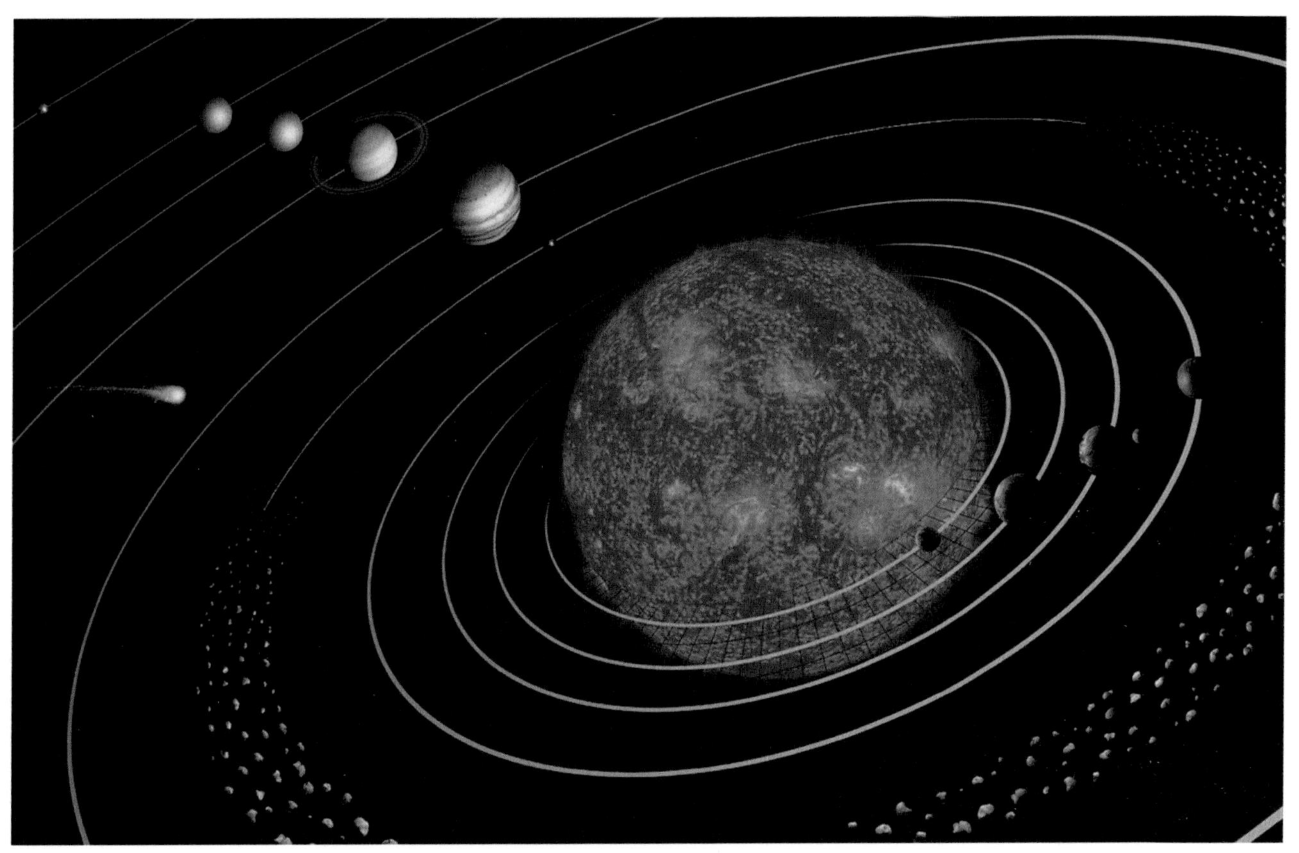

THE UNIVERSE IN A NEW LIGHT

CHAPTER 5

A Modern Meaning of Gravity

If a projectile were deprived of the force of gravity, it would not be deflected toward the earth but would go off in a straight line into the heavens and do so with uniform motion, provided that the resistance of the air were removed.

—Isaac Newton

More than a decade ago, the first gravity testing I recall I did—I did it solo. The experiment had to do with the diurnal variation of compasses. In the end, this simple compass or scope has been in use for centuries by travelers to find their way or navigate around this planet. It was noticed early on that a compass that remains in the same location will vary its direction by about one fourth of a degree on a daily basis. This is called "diurnal variation" (fluctuations that occur each day) and is attributed to our Sun's gravity because it happens on a 24-hour cycle. Our Moon does the same thing but with of course different timing and to much less effect. (1)

I built a compass using neodymium magnets stacked end to end to make about a 12-inch-long magnet that was a quarter-inch wide in the middle and tapered with narrower magnets on the ends. In the middle of this assembly, I glued a small round mirror so it was perpendicular to the magnets. I hung this assembly to a fishing-line harness inside a large wooden box that was put together with wood dowels and glue. No nails were used.

Also, I had a small pointing laser that I set up on the wall of our spare bedroom. It was about 10 feet away from the compass mirror that was placed on the air-mattress bed in the room. So what I had was a compass that was as far away from any metal structure in the bedroom as possible. The laser was pointed such that it hit the compass mirror and bounced the light back to the wall, near the laser.

I put a graphic on the wall that had a center zero-degree point with a quarter inch about equal to one fourth of a degree of compass movement horizontally.

One of the first things I noted was, the compass did not just move horizontally. It also moved up and down. It did move about one fourth of a degree every day, but not when I expected. It began moving in the early a.m. and finished back to the start about noontime.

Since I expected the compass to follow Earth's magnetic lines of force, I assumed that when it moved up and down, so was Earth's magnetic field, and the same should be true for the left/right changes.

The other thing I did not anticipate was the effect of the delivery vans coming to my house. The compass moved a lot more the quarter of a degree when they arrived and left my driveway. So I realized I had built a very sensitive system.

On one occasion, I was looking for unusual movement because Earth was in conjunction with star Lambda Aquarii, and I saw a reaction of nearly half a degree that was timed correctly—one of the many exciting moments of this research.

Finally, it was apparent the system was so sensitive that I needed to relocate to a desert location with no metal structure or traffic nearby. Needless to say, I have not done this yet.

This effort did make it real for me. I suspect the vertical movement from the Sun was because I do not live on the equator. I still think the Lambda Aquarii reading was real.

When I realized that Earth's magnetic fields are rotating with Earth, I could then imagine the magnetic fields coming into Earth from outside Earth—and they could be forced to curve in the same direction (eastward) by them. This, in turn, would yield compass effects earlier than the Sun position would dictate.

So this effort left me with no answers and some questions. If the diurnal variation of a compass is because of magnetism from the Sun, why is it before noon and in the direction Earth is rotating? If this diurnal variation is from gravity, why is it before noon and in the direction Earth is rotating. What else could this phenomenon be caused by? Could gravity and magnetism be synonymous—that is, the same thing?

After years of thought and discussion between my son and me, this just came to me in an epiphany. We had already developed the idea of two-dimensional matter as an explanation for magnetism and noted before, eddy currents, but now an obvious side effect of this two-dimensional magnetism when the magnet is not moving but any mass is moving nearby. Now I could see the opposite. The magnetic forces are moving in a common direction like a river of water and I am the object not moving.

This, in turn, results in a constant pressure on my whole body because it is affecting me at an atomic level where the eddy currents are. The ever-constant atomic-level collective force is why I weigh 170 pounds when I am at rest relative to the magnetic forces passing through me. I now needed to change perspective to see exactly how it happens. That means this is an understandable physical phenomenon that explains gravity.

Gravity Unsolved—or Resolved

Did you know, in all the scientific history of man, there has never been a conclusive explanation for gravity? Sure, there is plenty of information and math that describes what gravity does but nothing that adequately describes what it is or how it happens. Quantum mechanics science (a branch of physics that deals with matter and light on an atomic level) has invented the graviton as a starting point for the explanation, but it so far has failed to make sense to me.

There are other theories that are also obviously not the answer to *how* it happens. This was the case until now. Take a look at my nine-point checklist about how gravity may take place.

- ✓ Gravity is the unidirectional magnetic pressure due to the effects of two-dimensional, unidirectional magnetic currents passing through matter and inducing unidirectional eddy-current effects.
- ✓ OK, let's now assume that there are 10,000 stars close enough to Earth and the Sun to have significant Shaylik between them. All these stars are somewhat uniformly distributed around our Sun and Earth, and so there are 10,000 straight lines of significant Shaylik coming into both masses, right?
- ✓ And each of these lines of double-helix Shaylik has one of two chiralities. They all could have different frequencies. Only the Shaylik with the same frequency and chirality can combine through entanglement. The rest are repulsive and share a smaller and smaller area to exist in as they come into either the Sun or Earth. So most of the Shaylik become compressed between the other Shaylik coming into the masses. All of the lines of Shaylik are now nearly parallel, condensed, and coming from directly above.
- ✓ Then the Shaylik passes through me and I experience eddy-current forces from the encounter. The eddy-current forces are because the Shaylik is moving at nearly the speed of light relative to me and through me and toward Earth's center. As a result, we have the reason for gravity, and it is due to eddy currents.
- ✓ So now, no matter how or where you or I stand, our weight in pounds will not change. Lying down, standing up, at an angle, spinning, it will not change the fact that I weigh 170 pounds or whatever you weigh. This is because it is based on our mass alone. If nothing changes with our body organs, molecules, or atoms, we will still be in the same eddy-current environment.
- ✓ Also, the stars change very little in location over our lifetime. There is the occasional asteroid passing by Earth, and of course the Moon changes direction over 28 days.

- ✓ This thinking can likewise explain why the planets orbit the Sun. There is a balance between the centrifugal forces and the incoming-to-the-Sun Shaylik forces from all around the Sun and the eddy currents, which are also repulsive, forcing the planet toward the Sun.
- ✓ The simple notion that some invisible substance is passing through me, and you, without us knowing it is not new. If you study neutrinos (particles), you will find out that the current thinking is that about 100 trillion neutrinos pass through our bodies per second! This by itself, I expect, contributes to a little part of gravity on the sunny side of our planet. Neutrinos come from many sources outside our Solar System as well as our Sun. They also come from our nuclear power plants. In fact, the first detection of neutrinos was from a nuclear power plant.
- ✓ It's also acknowledged that magnetic forces from Earth's magnetism pass through our bodies with no apparent effect. The same is true for the fields of the magnet. I see the fields of the magnet as Shaylik that is chiral. The flow of Shaylik is there because it is a superconductor (a substance capable of becoming superconducting at low temperatures and able to conduct electricity without resistance), or at least acts like one, but it is different from the interstellar Shaylik of gravity. Our science cannot count the number of magnetic pieces passing through our bodies like the neutrinos are because these are lines of force, not pieces of force.

In conclusion, based on this checklist, the mysterious stuff passing through me, and you too, is not a new idea but is evidence it could happen. The only difference I see between the Shaylik I'm proposing and neutrinos is that the Shaylik is organized into two-dimensional magnetic lines, while it is not apparent that neutrinos are organized, but they do come in three types. Could that be one-, two-, and three-dimensional? It's possible. But the jury is still out.

Meanwhile, this is almost too obvious to mention, but magnetic levitation of frogs, magnets, plants, and water blobs using powerful magnets is a common demonstration of magnetic forces countering gravity. By the way, as far as I can tell, no plants or animals have been injured because of this phenomenon. If an invisible force can push a frog up against gravity, it can also push it down.

So there you have it. This is my biggest-of-all epiphany, which came to me a few years ago one night when gazing up at the stars and wondering about gravity and the galaxy (a big group of stars held together by gravity). It makes sense to me because it is logical and is a mechanical phenomenon. The key is being able to think in all three dimensions and understanding how eddy currents are an important force in the Universe.

NEW THOUGHTS, OLD UNIVERSE CLUES

- ✓ There is reason to believe this concept of gravity is reasonable and logical.
- ✓ One foundation of the theory is the well-known effects of eddy currents caused by magnetic lines of force.
- ✓ Gravity testing is connected to fluctuations that happen daily. And a compass can be a key tool but is sensitive to Earth's changes—natural or man-induced.
- ✓ There is a physical phenomenon that does indeed explain the foundation of gravity.
- ✓ A checklist of how gravity happens includes the Sun, Earth, Shaylik, and another factor!
- ✓ The eddy currents just may be the golden key force to the physical source of gravity in the Universe.

CHAPTER 6

Discovering Dimensions to Anyons

Whereas Nature does not admit more than three dimensions . . . it may justly seem very improper to talk solid . . . drawn into a fourth, fifth, sixth, or further dimensions.

—John Wallis

As I sit here early on an autumn morning, my thought is, there is only one possible explanation of the features of magnetic lines of force—one-dimensional and two-dimensional structures that are mass. Flash back to several years ago, when my son Jack and I came to this thinking while on the road in the motor home. We were on the way to our favorite California fishing spot, and we agreed that day about the magnetic-lines-of-force theory. Since then, we have continuously tried to find a reason that this cannot be true and, to date, have no reason to believe this assumption is false. So far as we know, no other human has ever defined what magnetic lines of force are; they just say what they do. This is also true for gravity: No human has previously explained how and why it works, just what it does.

From the first time I realized that Shaylik could be different dimensions, and they are one-, two-, and three-dimensional, I wondered why scientists took another path of thought. All I have seen on dimensions is that they assume there could be many more dimensions than just three. The many-dimensions and many-Universes thoughts are common in their writings. Why they didn't think about less than three dimensions is mind-boggling. I still recall Rod Serling, the narrator for the TV show *The Twilight Zone*, using the phrase, "There is a fifth dimension . . ." And each time I heard those words about science and superstition to man's knowledge and fear, it gave me pause. I doubt there is a fifth dimension in nature.

Dissecting Dimensions

The first epiphany was not recognized as that at the time and was the conclusion that magnetism could come in one, two, and even three dimensions. This conclusion came from logic only, and it took years for us to realize it was true. This happened during our annual fishing trip in 2013. The brainstorm baby that two-dimensional magnetism passing through me from above and holding me down due to eddy-current effects happened in my garage one morning. I do remember it was just before my birthday when the novel idea came to me, during pre-spring.

There's only one circumstance where I can find any mention of two-dimensional or one-dimensional mass. I'm glad I found it because it is the only information that adds credibility to this idea. Without these findings, I would have been way out on the logical limb.

Back in 2015, I first got the idea that magnetic lines of force are one-dimensional in their basic structure. This thought has been key since it first came to mind. In the past, I realized a lot of the features of magnetic lines could be explained by dimensionality. I also began to understand what we call "mass" could in reality exist in all three dimensions.

Mass is both a property of a physical body and a measure of its resistance to acceleration when a net force is applied. An object's mass also determines the strength of its gravitational effects on other bodies. The SI base unit of mass is the kilogram.

This definition doesn't exclude any idea of dimensionality. It only defines the properties it must exhibit. So there are many questions: a) Will one-, two-, and three-dimensional magnetism present these properties? b) Can magnetism be defined as a physical body? c) Can we observe its resistance to acceleration? d) Do the properties of all three forms of magnetism have the features of gravity?

The following are my thoughts on why the answer to each of these questions is a definite yes—no ifs, ands, or buts about my theory.

Did you catch the fact that I left out the word "attraction" in my group of questions about the features of gravity? Why? It's because I had already concluded that there is no such thing as "magnetic attraction"—only "magnetic repulsion." Yes, this is a paradigm shift that will be hard to swallow. Come along with me into an imaginary spacecraft and open your eyes to a novel idea (or two).

The Spaceship Full of Magnetic Pressure

The concept is simple if you look at it the way I do. So I imagine I'm snug as a bug, or as cozy as an astronaut can be, inside a pressurized spacecraft. It's a futuristic saucer-shaped ship like I used to watch in the sixties TV show *Lost in Space* or any present-day sci-fi films too. Envision, if you will, that there are small objects in my compartment that are floating all around. These can include water droplets, dust, Kleenex, gloves, and other things. All are stuff I can see floating around me and randomly distributed. Then, BAM! A meteor punctures my spacecraft in one spot! I now see myself and all the thingamajigs around me heading toward the hole. Is this attraction? It sure looks like it. Of course it is NOT attraction.

But what I didn't tell you is that, in this case, I am a monkey in space. I don't know about air and what it can do as it is escaping my spacecraft. So all I know is, everything that isn't attached to the spacecraft is being attracted to the hole. The forces involved are invisible and I don't understand what is really happening. Remember, I'm a primate in a spacecraft.

What is really happening is, every atom, molecule, dust particle, glove, water droplet, spacecraft walls, etcetera, and I are repelled from each other by the magnetic forces of our atoms. This is the reason that our shoes do not pass through the floor we are standing on. Atoms are mostly nothing. The distance between an electron and its nucleus is huge and is empty. Except, the magnetic fields of that are ever present everywhere, from everything. Your shoes do not pass through the void between atoms, electrons, and nucleus because it, your shoes, and the floor are not a void in terms of magnetic forces and they repel other magnetic forces. This is only noticeable when three-dimensional matter gets close enough to other three-dimensional matter for them to repel each other in accordance with the inverse square law. As they approach each other, the two masses will increase the repulsion by four times every time the distance is halved. Eventually the forces become enough to keep them from any physical contact and a balance of forces is achieved.

So a pressurized vessel, like a spacecraft, can be full of magnetic pressure. When that pressure is released through a small hole, the magnetic pressure everywhere decreases because there is less and less matter related to magnetic fields. And it's this thinking that can also explain why the planets orbit the Sun. There is a balance between the centrifugal forces and the incoming-to-the-Sun Shaylik forces from all around the Sun and the eddy currents, which are also repulsive and flowing into and through the planets, forcing the planet toward the Sun.

More Stuff on Dimensions . . . and Anyons

Now, back to dimensions. Some scientists indeed believe that one-dimensional and two-dimensional structures do exist. They call them "anyons." Interestingly, one of these anyons they believe could actually be used in a new type of computer processor. In fact, not only do physicists prove anyons are real—but we get a bird's-eye view of a third kingdom of quasiparticles that only arise in two dimensions. (1) (2)

So if scientists believe anyons can be used in a quantum computer, then by definition they think they have detectable features that can be changed or manipulated. Plus, it can affect other mass too. This, in result, means one- or two-dimensional particles do exist and they have at least one quality of mass. Go ahead—read on.

Easy as 1-2-3-Dimensional Matter

I've already told you about one-dimensional Shaylik and how it would interact with three-dimensional matter. Here are more ideas to chew on while I dish out outside-in thinking on this multiple-dimensional concept.

1. *A One-Dimensional Matter:* I talked about it as a one-dimensional rod smaller than an atom that I was pushing into sand with no or little resistance. Well, I actually don't think this is a by-the-textbook-definition situation. For something to have no size in two dimensions isn't

logical. It could not exist. So it has to have some size in the other dimensions, but it's just so small in size that it won't react with three-dimensional matter except under, perhaps, certain circumstances.

2. *A Two-Dimensional Matter:* Let's look at two-dimensional matter and consider what it would be like in a three-dimensional world or environment. Think of a piece of sheet metal about one-foot square. This sheet metal has such a small thickness that it will pass through and between the parts of the atom. It still has a dimension of thickness, but it's just too small to measure. Now put this sheet of metal in water so it is facing you. That is, you are on its side. Now try to move the sheet up and down in a barrel of water. Up and down is easy to do, right? Next, try to move it left and right—again it is easy to do this. Then move it away from you or toward you. Ah, now there is a lot of resistance to my (or your) efforts to move it. It's just like a boat paddle works in only one way to propel the boat efficiently. Two-dimensional magnetic lines of force would act the same way when it is in a magnetic environment or a three-dimensional environment. Forces would be greater in the three-dimensional environment, but the basic features would be the same: Most resistance is when it is moving sideways.
3. *A Three-Dimensional Matter:* So do the three types of Shaylik—one-, two-, and three-dimensional—qualify as matter? We know three-dimensional Shaylik does because it is matter. We know that what is holding the Universe together is called Dark Matter, and now we know what that Dark Matter is: two-dimensional magnetic lines of force. So it is matter in all three of its forms, or a three-dimensional matter.

Braid Groups and Anyons

Another field of intense investigation involving "braid groups" and related topologic concepts in the context of quantum physics is in the theory and (conjectured) experimental implementation of so-called anyons. Nope, I'm not talking about ponytail braids—this is a physics topic.

There's research describing interesting features of two-dimension anyons. And they're in what's called the "braid group mathematically." This means to me that the math says they can combine into braids and even the double-helix.

Another example:

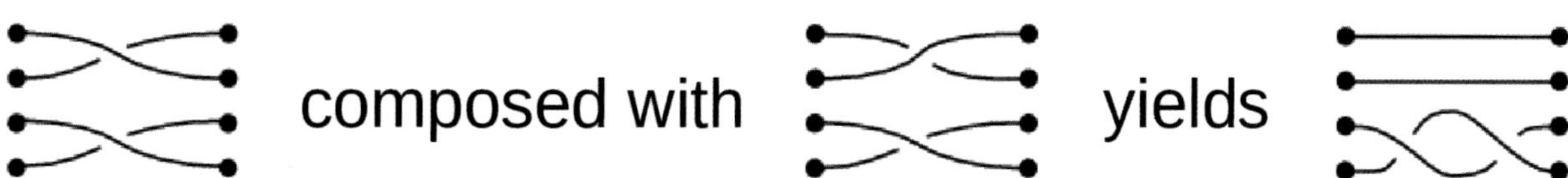

Open Sourced - Wikipedia

The lower-right assembly example is the double-helix. The double-helix is a basic large- and small-scale structure in our Universe as proposed herein.

Remember the red-and-blue rope that I assembled from just two types of Shaylik? That is a braid or braiding. The anyons they are talking about can be made into a rope or tube or whatever, just like Shaylik. (If lost because of the scientific-ese, go back to chapter 2, "10 Puzzle Parts to Legos and Shaylik.")

Research shows, in mathematics, the braid group on n strands, aka the Artin braid group, is the group whose elements are like classes of n-braids, and whose job is composition of alignment of braids. A sample of braid groups include "knot theory," where any knot may be represented as the closure of certain braids—it's all part of mathematical physics. But that's another book. (2)

Dimension Definitions

https://www.youtube.com/watch?v=ylz5_5wSMoc

View from 3:40 to 4:50.

The above link is of a presentation by Andreas Blass for the Simons Institute. In it, he tells of two definitions of two-dimensional matter: 1) a sheet of paper with no thickness; and 2) a magnetic cylinder that is infinitely long. The second example is very similar to what I propose as two-dimensional Shaylik.

The logic seems to be if, mentally, I take two of the paper sheet dimensions and make them zero and make the one zero dimension infinite, I have a structure that is like a tube that is infinitely long and possibly very straight. Now I have a structure that has a location that can only be defined using two dimensions. I would call this structure the inverse of the paper sheet and it is the most common structure of Shaylik, which is invisible and can pass through normal matter with little resistance.

If I apply this same logic to a one-dimensional structure, that is take the known dimension and make it zero and take the zero dimensions and make them infinite, then I have a sphere. So the inverse of a ten-inch-long string that has no diameter, and is then by definition one-dimensional, is a sphere because it now has two infinite dimensions due to its continuous surface and one zero dimension, which is its location. This structure would also be invisible.

This is mind-boggling.

Anyons Are Superconductive

https://www.science.org/doi/epdf/10.1126/science.257.5075.1354

The link above is for AAAS members only and discusses as a fact that anyons are superconductive. This is the first evidence I have come across that confirms the thinking that Shaylik is a superconductor in its one-dimensional and two-dimensional forms. The article also says, "Anyons can be viewed as ordinary bosons or fermions that act as if they each carry a tube of magnetic flux around with them." This is another key thought about Shaylik: it can be between all masses big and small as a double-helix magnetic structure.

Bosons and fermions are the two classes of elementary particles of quantum mechanics. The fact that anyons are viewed as both suggests a common structural nature such as Shaylik.

The Bond Between Anyons and Shaylik

Yep, anyons are exciting because they are recognized by our scientists as real structures in our Universe. Plus, they are only one- or two-dimensional. Researchers also recognize, anyons are superconductors, which is a critical feature of Shaylik. So what the science gurus have discovered is Shaylik in a specific environment that happens to be subatomic (a particle that is smaller than an atom.) Furthermore, Shaylik is also a superconductor. So what scientists call anyons are key support for the idea of Shaylik.

Also, anyons are evidence that matter in our Universe can be one-dimensional and two-dimensional.

Because anyons follow braid-structure math, they can be in the form of the double-helix, a physical body. The evidence shows that a two-dimensional structure exists and can be in the form of a double-helix. The evidence also shows that like chiralities can interact while different chiralities react differently.

In addition, there's reason to believe that two-dimensional Shaylik is the recognized three-dimensional structure we call "magnetism." Further, the eddy-current effects mentioned in the previous chapters are a way to measure the effects of two-dimensional matter on three-dimensional matter.

So what does all of this mumbo jumbo about dimensions and anyons really mean anyhow? Anyons, and therefore Shaylik, have been found to be superconductors in their one-dimensional and two-dimensional forms. What's more, anyons, and ordinary matter as bosons and fermions, appear to be physically linked with a flux tube. These bosons and fermions are particles that scientists say make up ordinary matter. Flux tubes, possibly in the form of a double-helix, may exist at the subatomic level. The end result: Anyons, which follow braid statistics, may be like discovering the pot of gold to understanding Shaylik in all its forms.

The whole point of dimensions and anyons is that our scientists recognize Shaylik exists but only in the quantum-mechanics world of science. They see the dimensions as the mass and the superconductivity in one-dimensional and two-dimensional anyons, but they are missing the bigger picture. It adds more credibility to the concept of Shaylik.

Speaking of Shaylik, it doesn't stop here. Not a chance. Now that you are aware of the connection between anyons and Shaylik, it's time to dig deeper. In the next chapter, I share more keen observations that deserved to be looked at under the telescope and decoded—since there are always more questions to be answered.

NEW THOUGHTS, OLD UNIVERSE CLUES

- ✓ Magnetic lines of force have more than one-dimensional structures, and this fact is intriguing and opens doors to amazing stuff.
- ✓ Shaylik is one-dimensional but interacts with two-dimensional and three-dimensional matter.
- ✓ Forget the concept of magnetic attraction—only magnetic repulsion is the real deal.
- ✓ A spacecraft can be full of magnetic pressure, and when pressure is released, the magnetic pressure decreases like letting air out of a balloon. The answer is, because there is less matter related to magnetic fields.
- ✓ Anyons (and Shaylik) that follow braid-group stats may be the holy grail to understanding Shaylik and all its forms.

CHAPTER 7

More Surprising Shaylik Observations

Gravity explains the motions of the planets, but it cannot explain who sets the planets in motion.
—Isaac Newton

Truth be told, I've known for a long, long time that our planets' orbits are harmonically related. As a one-man science team, I did my own analysis and found a common multiplier for all our planets' orbital periods. The fact that they are nearly even multiples of each other is why sometimes most if not all the planets line up on one side of our Sun. I also researched the moons of Venus and Jupiter and found they have harmonically related orbits also. And sure enough, the result was what I had suspected all along. They all line up on one side of their planet on occasion.

What I didn't know, though, was if the minor planets were also in harmonious orbits with the planets. So I continued onward, searching for the answer. During my recent research on solar phenomena, I stumbled on data dated 08/22/2012 at 1746 UTC (Universal Time) that Jupiter, Ceres, Mercury, and of course the Sun were all in a line! It was mind-blowing. So at least one minor planet has an orbital period that is in harmony with the planets.

Shaylik Is Ba-a-ck for Harmony

I've had a theory about what causes this harmony since I imagined the Shaylik between planets and stars as a straight line to our Sun. I researched this subject and found that some scientists attribute this harmony phenomenon to gravitational interaction of the planets and moons of planets. What I'm thinking is, gravitational interaction is part of it, of course, but more to the point, how does this gravitational interaction happen?

The Shaylik between the planets and our Sun are straight and moving with the planets. The Shaylik between the moons are doing the same with their respective planets. Now consider Earth, for instance. Earth passes through the Shaylik from Mars to the Sun, and the Mars Shaylik is moving in the direction of all the planets with Earth. Earth is slowed just a little bit due to the eddy-current effect. It also may become statically charged during the passage through the Mars Shaylik. This is of course the two-dimensional Shaylik I'm talking about.

Now stay with me. This same thing can happen with the Shaylik from Jupiter. Jupiter is orbiting much slower than Mars, so when Earth passes through it, the effect will be shorter in duration and stronger than the Mars Shaylik because of the mass differences between the Mars Shaylik and the Jupiter Shaylik. This means there is no consistency between the two phenomena. Both will slow Earth, but the forces are different in duration and intensity. This would be true no matter what planet is affecting Earth or what planet may be affecting any other planet. This dance of the planets and moons has no metronome or drummer that keeps the cadence. Enter the drummer.

Twinkle, Twinkle, Stagnant Stars

The stars don't move at all compared to the planets and moons. They aren't stationary but don't move around our Sun much at all. What if there is a star or stars that the planets and moons come into conjunction with frequently? And what if this star or stars were closer to our Solar System and larger than any other star or stars?

There just happens to be such a star system. Alpha 1 and 2 Librae are two very massive stars that are relatively very close to our Solar System at 75 light-years away. These two stars are also near the ecliptic, which means it is likely all the planets and all the moons will on occasion be in conjunction with the Shaylik from this star system. This will not happen every orbit of every planet but will happen relatively frequently for all the planets in the same orbital plane.

Also, I don't know the diameter of the Shaylik from this star system. It likely is huge like the spacecraft-detected Shaylik in the following text. If that is the case, it's the drummer the planet dance needs to ensure it has harmony. But there's more . . .

Is Shaylik the Big Cheese of Our Universe?

One of the key features of Shaylik is that it takes the form of double-helix when it is between masses in Outer space. The first indication I observed that Shaylik could take the form of a double-helix was when I was observing what happened when ferrofluid (a fluid containing magnetic suspension like water) was exposed to a magnet. This concept was then confirmed through ACE spacecraft data and subsequently by *Science News* author Jamie Chambers's findings. This is a notion that I see as a key feature of two-dimensional Shaylik, which seems to be the dominant form, making up 80 to 90 percent of our Universe.

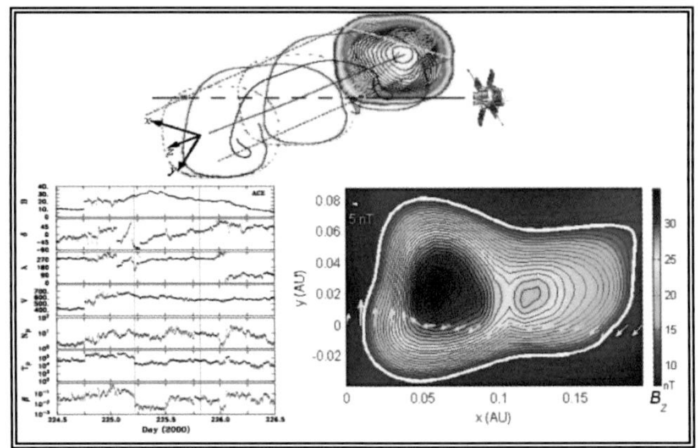

ACE News #66 – Nov 27, 2002 A double Flux-Rope Magnetic Cloud - NASA

The ACE news article (**http://www.srl.caltech.edu/ACE/ACENews/ACENews66.html**) tells of an invisible magnetic structure discovered by the ACE spacecraft that is in the form of a double-helix and is about 19 million miles across. This is the clearest example and evidence that the double-helix structure exists in the cosmos.

Features like this are critical when analyzing magnetic data from our Earth- or space-based magnetic detectors. This configuration will only yield single- or double-spike magnetometer data. The double-spike data will usually be two spikes that are not the same level of magnetic energy. This gives me an idea of what it is in the data I am looking for but have yet to discover the definitive answer.

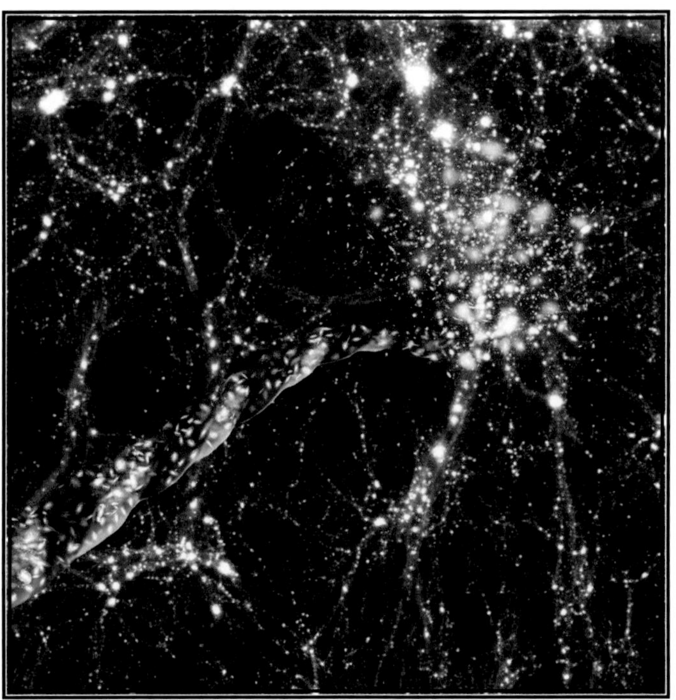

A. Khalatyan/J. Fohlmeister/AIP

Cosmic filaments are strands of dark matter and galaxies that rotate (illustrated). As the filaments spin, they pull matter into their orbit and toward galaxy clusters at each end.

This article (Jaime Chambers https://www.sciencenews.org/article/dark-matter-cosmic-filaments-biggest-spinning-objects-space) is about the largest structure we know of in our Universe, and it's shaped like a double-helix. It stretches millions of light-years and contains most of the mass we can see.

When I view this graphic, I imagine there is an invisible double-helix surrounding a visible plasma that is being shepherded by the magnetic lines of the helix.

Additional evidence of the double-helix structure at microscopic levels is found in a 1400X microscope picture that was amazing to me. The image shows a double-helix or chain structure of the magnetic lines of force. The viewer needs to understand this is an oil mixed with iron, almost nanoparticles, being affected by a nearby permanent magnet. This oil mixture is between two very flat pieces of glass. Look closely and you will see the magnetic features are appearing and disappearing through the oil mixture. You can view this picture at the following site: https://www.ferrocell.us/intro.html. Pan down to find the green almost-square picture noted at 1400x image.

So far in this book, I've shown the possibility of the double helix magnetic structure at all scales. This data supports the idea of a very stable structure made from Shaylik that has a repeating fractal nature that also may be found in a quantized size.

Meanwhile, like other interesting theories, more research is needed before we have definite answers to show the proof is in the magnetic energy. While on the topic of magnetic energy, what about electromagnetic radiation? In chapter 8, it's time to decode this form of light and show that it matters.

NEW THOUGHTS, OLD UNIVERSE CLUES

- ✓ The two-dimensional Shaylik exists from smaller parts of an atom scale on up to bigger galaxies.
- ✓ It seems reasonable the double-helix Shaylik could be smaller than the parts of an atom.
- ✓ In many cases, if not all, the double-helix structure is rotating.
- ✓ The double-helix Shaylik has both ends ending at three-dimensional masses.

CHAPTER 8

A New Look at X-Rays

She is the child of the Universe.
The Universe makes rather an indifferent parent, I am afraid.
—Charles Dickens

One afternoon back in Long Beach, California, I experienced an amusing and memorable dental appointment. The dental assistant in a white uniform was taking X-rays of my teeth. (I now know these are radiographs using a burst of X-ray radiation to find hidden structures.) During the uncomfortable procedure, I raised my hand up for her to stop probing my mouth with the hard white squares.

"Did you hear about the BIG X-ray discovery?" I asked. After all, it was a big deal in 1962.

She nodded no while holding on to the square radiograph to put inside my mouth.

"It's huge!" I exclaimed. "Scientists used the first X-ray telescope in a space rocket! Imagine the possibilities of it!"

She chuckled, sighed, and shrugged her shoulders, as she was on her own mission to do her job. I heard the words. "Open wide." I complied but didn't understand why this X-ray lady didn't share my fervor as did scientists worldwide. But X-ray findings didn't stop that day, and groundbreaking research continues.

Shining Light(ning) on X-Rays

X-rays are one type of light designated by its frequency. The frequency of X-rays is in the highest of all the frequencies. These phenomena we call light are electromagnetic energy. This energy exists all across the electromagnetic spectrum, from as low as one cycle per second to 50,000,000,000,000,000,000,000 cycles per second. Now that's a lot of zeros!

Lightning is familiar to us all as "white light," but it's much more than that streak in the sky that happens before the sound of thunder. Lightning, in Earth's atmosphere, puts out electromagnetic waves from less than one cycle per second to two X-ray levels and possibly beyond. Ask anyone listening to an

AM radio during a thunderstorm. Noise flashes in the radio no matter what frequency you are listening to. Try it during a storm and you can hear it for yourself!

So why am I telling you all this mumbo-jumbo science about electromagnetism?

Simply put, the point is that X-rays, like any light, have a key feature or features that are important to this chapter. All electromagnetic waves can be absorbed, emitted, or reflected by three-dimensional mass. That means, in any photo of X-rays, if there is three-dimensional mass that either absorbs or reflects X-rays, the picture would show a change in the amount of light in the X-ray spectrum. If it's emitted, it would be brighter than the rest of the photo. If it's absorbed or reflected, it would be dimmer than the rest of the photo.

Also, there's scientific data that shows the size of something if it absorbs or reflects electromagnetic energy at certain frequencies. Therefore, if an invisible something (whether it is an object or an alien from Mars) absorbs or reflects X-rays, its physical size can be estimated. If this invisible something is Shaylik, then information about its basic size could be derived.

This may be the only way of deriving the basic structure and size of one- or two-dimensional Shaylik, which is for now an unknown.

More Light on X-Rays, Please

So yes, X-rays are an interesting form of light. Very interesting indeed, and very energetic compared to the light our eyes detect, and with very short wavelengths. Longer-wavelength visible light can barely penetrate our skin, while X-rays can pass through our whole bodies. They do not pass through without some resistance, and that is why we can get what amounts to black-and-white pictures of what is inside our bodies. Just like visible light, X-rays can hurt our bodies if we get too much exposure.

We typically produce X-rays by smashing electrons at a high speed into some metal that's in a vacuum tube or chamber. The energy or speed of the collision and the type of metal will produce various frequencies of X-rays. And nature does indeed produce X-rays also through lightning strikes—not just while you're sitting in a chair in a dental office.

There are many sources of X-rays coming from stars and galaxies. But I want to focus on X-rays produced by lightning. When there is a lightning strike in our atmosphere here on Earth, the gases in our atmosphere are ionized by the great energy pulse from it. This causes some atoms to emit X-rays as they try to get back to a stable form. Ionized gases are called "plasmas" and are abundant on our Sun, which also emits X-rays from its hottest areas.

So it's obvious that nature is extraordinarily good at producing X-rays where there is sufficient matter and energy.

The two-dimensional Shaylik I've described, time after time, to you is energy because it's also mass. Mass and energy are interchangeable. It's entirely possible, at times with sufficient energy in this Shaylik, that when it encounters a plasma or just a gas that is excited by the Shaylik, X-rays will be produced. This would make the presence of Shaylik apparent to an X-ray telescope. I'll be watching for indications this is happening, and perhaps you can too. If it does, it will appear as a straight line of X-ray energy. Keep your eyes (and mind) wide open to this incredible happening.

Missing the Mark of X-Ray Data

Not unlike the dental assistant who focused on getting pictures of my teeth and lacked my enthusiasm about the 1962 discovery of astronomical X-rays, I'm sometimes amazed that our scientists ignore something that is staring them in the face. This is one of those times.

I have searched off and on to discover an explanation of the phenomena discussed in this chapter. Alas, there's no scientific paper where I can find anything that even mentions the obvious dark lines in present-day X-ray data. There are other X-ray chasers (think of sci-fi films with passionate crews who follow the trigger of a happening) asking the same questions I do. But there's still no reasonable answers yet; however, that doesn't mean one day the missing puzzle part will be found by scientists.

Meanwhile, there are online graphical displays of the galaxy around us at various frequencies of light. There's a viewing bar that lets you look from radio waves to gamma rays by moving the bar. Since the display is of the sphere around us and it's in two dimensions, the view is nonlinear or distorted. It's like a map of Earth shown in two dimensions. The middle is a different distance scale than the edges, in order to show the sphere in two dimensions. Structures are shown closer together on the edges than they would be if they were in the middle. If there is something out there that is a straight line—and it's not coming from directly in the middle of the graphic—the straight line will appear curved even though it's not actually curved.

And this is exactly what happened in the X-ray view. There are straight lines coming from the Large Magellanic Cloud (LMC), aka a satellite galaxy of the Milky Way, which is near our galaxy and not shown in the center of the display. So the lines, while actually straight, are shown curving. The point? It's that these lines exist with no explanation whatsoever for them that I can find in my research. Again, the verdict is still out, and more scientists need to look into this matter. (1)

If you look closely, it appears the curved dark lines are coming from the Large Magellanic Cloud (LMC), which is a companion to our galaxy. So if this is double-helix Shaylik, it's coming from the three-dimensional mass of the LMC and going to some other three-dimensional mass like Earth, or at least our Sun.

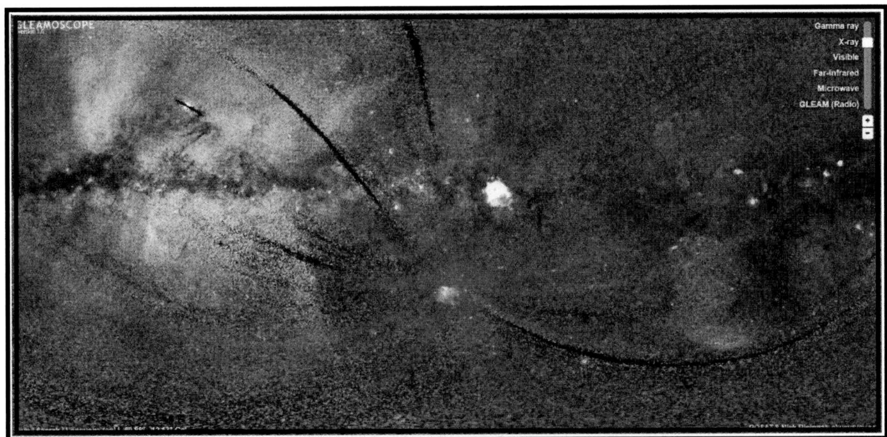

Authors screen print – ROSAT X-ray data

A possible explanation? It's feasible it has to do with the design of the cameras, which I doubt, considering the other frequencies not showing these lines. My justification is, this is Shaylik coming from the Large Magellanic Cloud that is passing by Earth and on to a star, possibly our Sun! The Shaylik must be either reflecting the background X-rays or absorbing the X-ray light. Also, the reason these lines are not all over the graphic is because the effect can *only* be seen if our perspective is almost directly in line with the phenomenon. This, in turn, means only when the viewer is at one end of the line will it become apparent in the X-ray picture.

If I Had an X-Ray Telescope

Remember the "If I Had a Hammer" protest song, a hit tune for Peter, Paul and Mary in 1962? Well, for me, the title should be changed to "If I Had an X-ray Telescope." Seriously, if I had access to an X-ray telescope, I would take many short-duration pictures of a planet like Jupiter (the fifth planet from the Sun and the largest planet in the Solar System) before, during, and after an Earth conjunction, to prove or disprove this notion.

Hubble Space Telescope Gets Picture of Small Spiral Galaxy!

At the end of 2021, excited astronomers using the NASA/ESA Hubble Space Telescope captured a photo of the small spiral galaxy NGC 1317. It's astronomical discoveries like this that call for a top-notch telescope and lab giving scientists data across the electromagnetic spectrum. And this is what I'd love to have in my Shaylik research repertoire.

This is the only evidence I have to date that suggests that Shaylik can be detected with X-ray cameras. I suspect that Shaylik could be like a gas cloud (also called an "interstellar cloud," which is an accumulation of dust, gas, and plasma in our galaxy and in Outer space). When a gas cloud is cold and has a light source behind it, we can look at the spectrum of light coming through the cloud and tell what molecules or atoms it is made of. This is one of our best tools for analyzing stars, gas clouds, and whole galaxies. In the case of the cold gas cloud, the data is called an "absorption spectra." When a gas cloud is excited to higher energies by a nearby star, it will glow, and this is called an "emission spectrum." The cloud emits its own light that can be analyzed to find out the molecules or atoms it is made of. This has been a great and beneficial tool for many years.

The bottom line: I suspect that Shaylik will act in the same way as the gas cloud. If it's cold with a light source behind it, there'll be dark X-ray absorption lines. If it's excited to higher energies by large amounts of energy passing through it, there'll be X-ray emissions from it. In any case, I feel (blame it on a gut instinct) it would be beneficial to research this possibility. And yes, there is reason to believe that X-rays are the only indicator of the presence of Shaylik. That is, as well as gravity, with respect to Newton.

One more thing. By observing the sky with X-ray devices—yes, these are exciting instruments—scientists gather vital data to help find missing puzzle parts to how the Universe began and how it evolves, and discover its ultimate destiny. (2)

NEW THOUGHTS, OLD UNIVERSE CLUES

- ✓ X-rays are used in medicine but are also a valuable device put to work in Outer space since the sixties.
- ✓ There are a variety of causes of X-rays, and lightning in the sky is just one of them.
- ✓ In 1962, X-rays were discovered, and more discoveries followed, paving the path of usage of them and planetary science . . .
- ✓ The first rocket flight to find a cosmic source of X-ray emissions was launched by a group at American Science and Engineering. This rocket flight used a small X-ray detector. It found a bright source they coined "Scorpius X-1."
- ✓ An invisible object to aliens (friendly or hostile) that absorbs or reflects X-rays may be detected—and Shaylik isn't excluded.
- ✓ Nature is good at producing X-rays where there's matter and energy.

PART 4

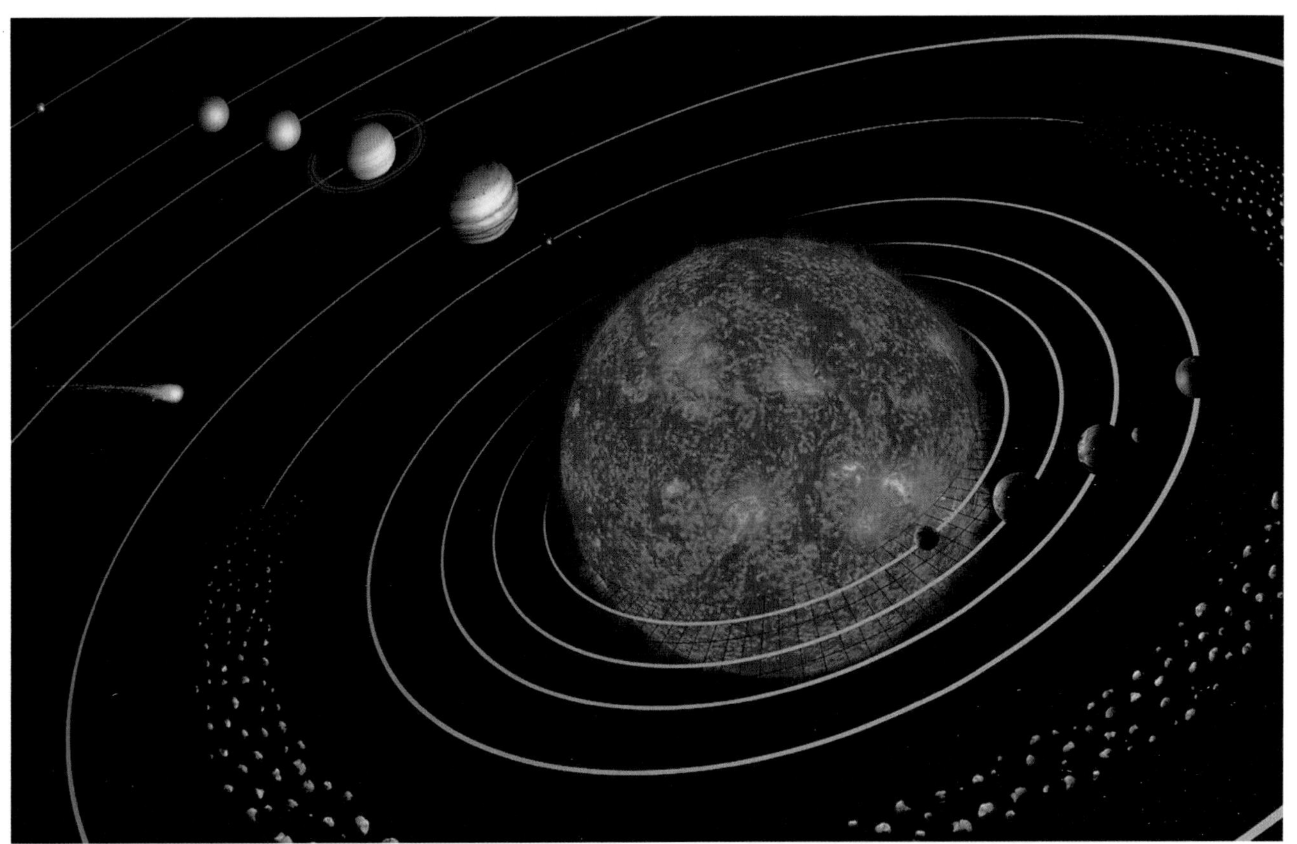

BUILDING BLOCKS TO A SOLAR SYSTEM

CHAPTER 9

Earth's Magnetic Force

The magnetic force is animate, or imitates a soul;
in many respects it surpasses the human soul
while it is united to an organic body.
—William Gilbert

About 20 years ago, I watched the sci-fi film *The Core*. The scenario is, for unknown reasons, Earth's core stops rotating. The result: It causes the planet's electromagnetic field to quickly cause chaos. When Earth's inner core stops rotating, it causes the planet's electromagnetic field to quickly fail. The consequences are dire. Birds lose their ability to navigate (this can affect anything from car collisions to aircraft crashes), and extreme weather and electrical superstorms occur and wreak havoc on Earth.

Sure, it was just a movie, right? Wrong. Our Earth's magnetic force is real and essential to life or death. What are magnetic fields anyhow? Scientists will tell you it is a term called "the force." The permanent magnet generates force called "magnetic field" when it moves toward the nickel, cobalt, and iron. They are considered the ferromagnetic materials. Magnetic fields are vital in the development of technology and much more, as I saw in the movie. (1)

Entering the Land of Magnetic Fields

Did you know we have magnetometers placed all over Earth's surface? These devices are used to measure the magnetic field, especially its strength and orientation. A compass, for example, is a simple one and is used to measure Earth's magnetic field. In fact, the first accurate magnetometer was built in 1833. Now I have one in my telephone. These Earth-based magnetometers are mostly in the northern hemisphere and are used for recording the changes to the intensity at any moment at a given place. The output of most of these magnetometers is available in near-real time. (2)

I'm hardly alone in discovering the usefulness of magnetometers. These gadgets are in Isaac Newton's data. What's more, the force (nT) is named after the scientist. Force is only one feature of natural magnetic lines. Another measurable feature is frequency. There is such a thing as static magnetic forces, and they are associated with the magnet and the direct-current electromagnet. Man-made magnetic forces, as far as we can tell, have no frequency. The naturally made magnetic lines of Earth have a frequency of 1.3 kHz (the hertz symbol is the unit of frequency, named after Heinrich Rudolf Hertz, who found proof of electromagnetic waves) at Earth's equator and 2.5 kHz at or near Earth's poles. These are audio frequencies, which may explain why my ears ring once in a while.

I imagine what's happening at times is, the magnetic forces in the form of a double-helix are so large that I only catch the edges of the spiral as the Shaylik helix is passing by and through me. The spiral edges are cyclically passing by and through me, causing the magnetic pulses we detect or hear as a frequency.

If this concept is correct and Shaylik is moving at the speed of light, then it makes sense that I should be able to calculate the turns per mile of the helical structure. Just casual observation of the numbers tells me these helixes are very tightly twisted. Two thousand cycles per second (2 kHz), and light travels at 186,000 miles per second. 186,000 divided by 2,000 equals 93 miles between each spiral turn. Hum, I have seen a number near this in studies of the harmony in our Solar System planet orbits compared to the orbital periods of electrons in hydrogen atoms. I came up with a common factor for both: 91.314 days of time for our planets and 91.175 nanometers wavelength for the hydrogen atom's various wavelength features.

How does time relate to distance in nanometers (one billionth of a meter)? They both are a measure of wavelengths. The planets' orbital periods are time-based cycles per unit time, which is equivalent to wavelengths. The use of nanometers is simply a way to express wavelength size, which means both numbers are related in terms of wavelengths.

Studying Shaylik . . . and Magnetic Force

No, I don't expect the Shaylik around our planet to be the same frequencies as the Shaylik between planets or stars. Our Sun has magnetic lines of force similar to Earth's. I also don't expect the Sun's magnetic line frequencies to be the same as it is on Earth or the Shaylik between planets and stars. Also, I doubt the star-to-star frequencies to be the same as the planet-to-planet frequencies. All these circumstances are physically incompatible frequency circumstances. There probably is some compatibility of frequencies within a group. That is star to star, planet to planet, and galaxy to galaxy. Most of the time, the Shaylik flowing around out there in space is incompatible with the other Shaylik it may encounter by chance. This is one reason I think magnetic forces are only repulsive—and do not attract.

Mother Earth and the Magnetic Fields

Back on planet Earth, we know that the magnetic fields are generated for the most part by triboelectric effects (a charge of electricity caused by magnetic friction, also called eddy currents). I see triboelectric results as simply what happens when two masses are nearby each other and are moving at different speeds. This is happening all the time throughout the cosmos.

Here on Earth, the major player is Earth's core spinning inside a layer just outside it that is spinning at a different speed. The magnetics of the core are interfacing with the magnets of the next layer out, and a magnetic field is produced. These magnetic fields are somewhat aligned with each other and are 90 degrees to the angle of the motion. So the spinning core is turning horizontally just as Earth's surface is, and the magnetic lines created turboelectrically are vertical and oriented in a north-south direction. Think of an engine such as a turbo-propeller jet. These north-south magnetic lines are also spinning in an easterly direction with Earth's rotation.

As a result, this is exactly what the magnetometers are measuring at Earth's surface. These magnetic lines extend out from Earth's surface infinitely and collide with the Sun's similar magnetic lines of force and the solar wind. This, in turn, means the Sun's effects on Earth's magnetic lines are relatively constant except when there is a coronal mass ejection hitting Earth. Then there is lots of activity extending from the poles on down toward the equator.

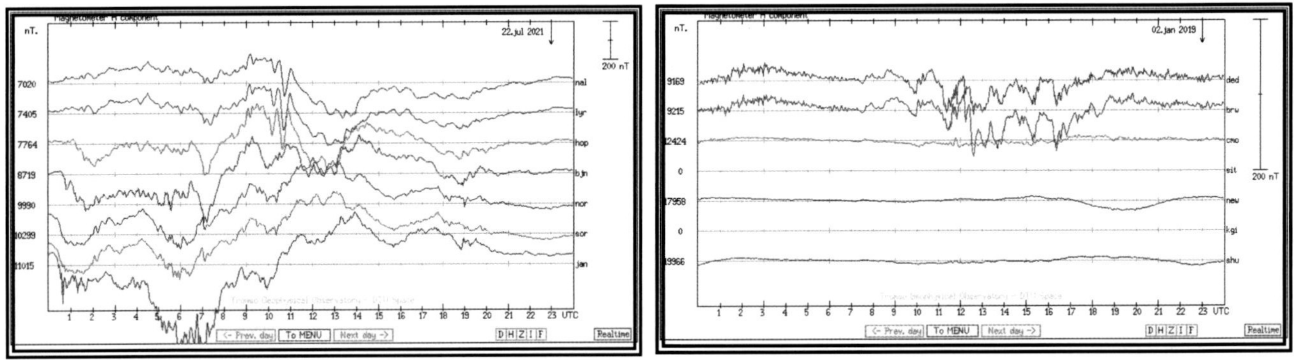

National Oceanic & Atmospheric Administration (NOAA) – spaceweatherlive

The Plot of a Stackplot

The graphics aren't during an Earth/Star conjunction and are simply examples of two significantly different situations. The North American Stackplot (a plot in this region where magnetometers detect a geomagnetic disturbance and react) was when the Sun's effects on Earth's magnetic fields were minimal. Note the right vertical bar shows the range for 200 nT. On the European Stackplot, this vertical bar is very short for the same 200 nT forces. The bar is four times longer on the right graphic than the left. This is notable to understand because the energy levels are four times higher on the left graphic, which was during a coronal mass ejection (CME). Its strong eruptions near the surface of the Sun, triggered by twists in the solar magnetic field, can cause a ripple effect through the Solar System—and they put satellites to power grids (a network of electrical transmission lines connecting generating stations to loads over a wide area) on Earth in harm's way. It's also important because the energy levels I'm looking for are small and wouldn't show up as well on the left, less-sensitive phenomena.

So what I'm seeking to find is either a single or a double spike that is either up or down and at the moment of a conjunction of Earth and a nearby star. The reason I am looking for only single or double spikes is because this is what I would get if Earth was crossing laterally through a double-helix structure. Most of the larger nearby stars are named, so I'm typically gazing at a named star to give me the results I am studying. These spikes I'm watching for will typically not be obvious during high activity from our Sun; mostly during solar calm will I (or you) see this happen.

Earth is small compared to the size of the double-helix flux rope and may take hours or days to pass through them. So far, my experience has been hours phenomena more than days. Also, if the frequency of Shaylik we pass through is between 1.3 and 2.5 kHz, it will show up anywhere between Earth's equator and its polar regions. Example data follows:

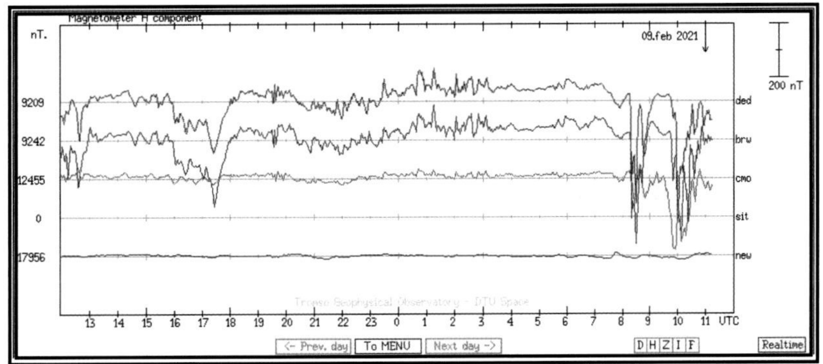

National Oceanic & Atmospheric Administration (NOAA) – spaceweatherlive

On 02/09/21 at 0230 hours local time (0930 UTC is between the two right down spikes), Earth was in conjunction with the star TYC 827-13-1 83ly away and -0.2 degree ecliptic. This star is located at 09h 33m 08s latitude. Note the classic double spike appearance. The nT sensitivity is normally low at this time in the Sun's solar cycle. Also note the two spike peaks are about 1.5 hours apart.

I collected data like this example from 08/28/2020 to 05/06/2021. The results weren't what I had hoped for. Why not? In the times I got a correlation with a conjunction, they weren't beyond random chance readings and they didn't repeat in previous years, at least not significantly. I did notice, however, a variance in location of these data in line with my thinking that different frequencies are involved that would show up at different latitudes on Earth.

Later, I found that Mercury (the smallest planet in our Solar System and the closest planet to the Sun) conjunctions with stars, and X-ray flares on our Sun could yield the data I was looking for, which supports the conjunction-effect concepts for Shaylik. I still believe this approach could yield valuable results, but either I'm missing a puzzle part or I'm making some incorrect assumptions. Entering unchartered space is how we discover new things, right?

So what we have here is a well-documented failure to support my hypothesis. The readings have varied in frequency based on the location of the data. Meanwhile, I haven't let go of this as a supportive data source, but it seems we'll have to wait about 10 years for the Sun to be at a minimum again.

As I put these musings on the side, it's time to dish up another topic not unknown to mankind. We have several known planets—but mighty Mercury is one that has caught my attention. Read on—find out why this sphere is worthy of its own chapter.

NEW THOUGHTS, OLD UNIVERSE CLUES

- ✓ Magnetic fields greatly affect our planet.
- ✓ Magnetometers are devices to detect changes in Earth's magnetic field, which helps scientists monitor Earth changes.
- ✓ Shaylik from Earth probably doesn't have the same frequencies as the Shaylik between planets or stars.
- ✓ The Sun's effects on Earth's magnetic lines are nearly constant—except if a CME hits our planet. Then all bets are off, more or less.
- ✓ Mercury, its conjunction with stars and nearly simultaneous X-rays from solar flares could yield proof to show the conjunction-effect concepts for Shaylik. Time will tell.

CHAPTER 10

Say Hello to Planet Mercury

I had rather be Mercury, the smallest among seven [planets], revolving round the sun, than the first five [moons] revolving around Saturn.
—Johann Wolfgang von Goethe

Welcome to planet Mercury. This littlest planet and its importance spawned a new idea that came to me one day while I was studying the different planets. The notion is that Mercury could be the Rosetta Stone for information on why and how solar flares (an intense eruption of electromagnetic radiation in the Sun's atmosphere) happen on our Sun. In other words, this small planet could be the key to further understanding how exactly some things happen on our Sun.

Because it's so close to the Sun, any life on Mercury would be impossible due to the excessive heat. During the daytime, the Sun would appear three times larger and more than 10 times brighter than it does here on Earth. All of that sunlight can push temperatures as high as 800 degrees F—uninhabitable. It would be like living inside a microwave. On the back side, there's no atmosphere to maintain the heat, so temperatures can plummet to -300 degrees F—colder than Antarctica!

Go Ask Alice in the Rabbit Hole

Recently, I've noticed that sometimes a conjunction of our Sun, Mercury, and a nearby named star correlates with the timing of a solar flare on our Sun. I can see the moment of this direct line of the Sun, Mercury, and star using astronomy software. As noted earlier, this software allows the user to look at the location of planets and stars from the perspective of the Sun. And this is how to construct a time and date for the astronomical alignment. It's as if I were standing on top of the Sun with my telescope like Alice viewing a new world when she falls down the rabbit hole.

At times, I have looked at historical data for support of my outside-in thinking theories. I've noted that the raw data coming to our scientists is rarely available to the mainstream audience at no cost. I also understand that there's a lot of data in the raw form that will be lost when that data is processed for public consumption.

Coronal Mass Ejection Phenomena

Most CMEs last more than an hour and can be represented accurately in a reduced data format. What I'm seeking can happen in minutes to days. This situation makes me very skeptical of processed data. Think of junk food versus real food grown on a farm. That's why I depend on real- or near-real-time data in most of my studies. Historical phenomena confirmation would be nice, but it's not required or even looked at by me until I'm sure the real- or near-real-time data is supportive of my findings.

Years ago, I realized change in all things is universal. It's the one thing I can depend on throughout my life like the ocean—it's forever changing in depth to its ebb and flow or natural disasters like coastal flooding or a tsunami. This is also true for our Universe. Nothing will stay the same forever—and this fact includes the evolutionary Universe. When I first started getting this mercury data, I was getting confirmations weekly. Now I'm getting almost nothing.

Mercury and Solar Flares

As scientists claim, gravity is a very weak force compared to the other forces out there on Earth and in Outer space. So the information I'm looking for won't shine, but it's masked by other forces most of the time. Also, I have just begun investigating the Mercury connection with solar flares—but have little to rave about as of yet. But one notable thing has become obvious in that the location of Mercury doesn't seem to correlate well with the location of the flare on the Sun. So there is that.

The lineup of Shaylik and Mercury is intriguing because of its effects. Because I believe Shaylik is less likely to be slowed from traveling at the speed of light than visible light itself, I suspect solar flares will happen slightly before a conjunction. This is because the lineup is based on visible light reaching us where the flare is probably caused by the faster Shaylik. In fact, the flares may be latter than the conjunctions because the inertia of the Sun's plasma must first be overcome. Again time will tell.

NOTE: Most of the graphics in this book are not photographs. The resolution is poor compared to photographs the reader is used to seeing in a book. They are screen shots from my computer that the user can reproduce for free in the internet. Just enter the subject date and or time to verify what I am pointing out with the screen shots. The AstroGrav program is the only program that must be downloaded for a fee. All the rest are free and on the web. I expect some to question what I am saying and if they have the time will verify the data through their own online investigation.

Some of the initial data I have gotten relating to Mercury suggests that the solar flares from Mercury conjunctions happen after the moment of peak conjunction. On a Thursday in late October 2021, I was anticipating some solar flare activity may happen that day. I looked ahead in time with the AstroGrav application to see that Mercury was in conjunction with the star Pollux in the afternoon. Two days later, on Saturday the twenty-sixth, Mercury was going to be set up in conjunction with the star Ψ CNC. The following data is what I observed.

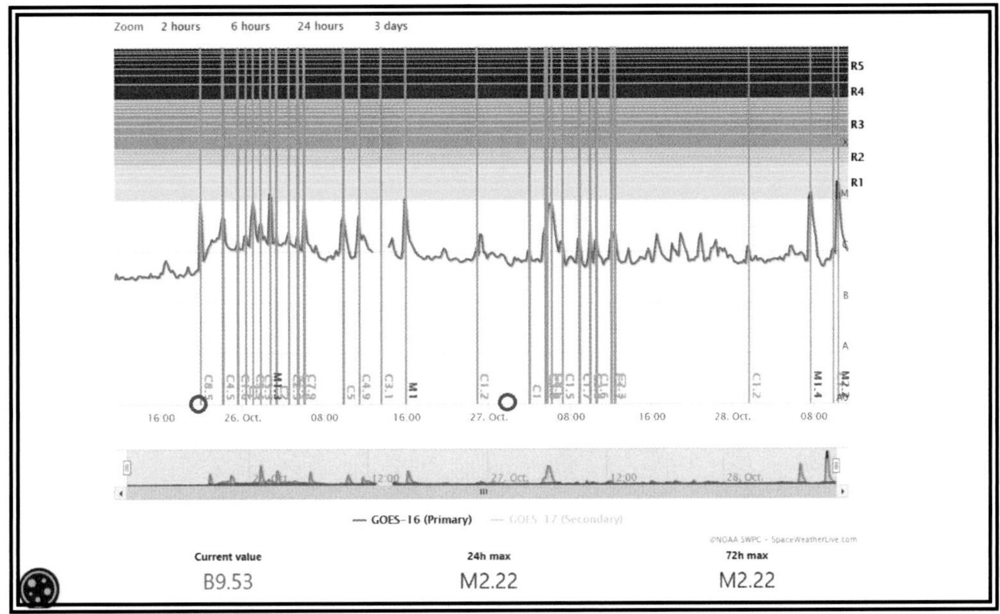

National Oceanic & Atmospheric Administration (NOAA) – spaceweatherlive

- The two blue circles are the moments of conjunctions of stars Pollux and ȝ CNC with Mercury.
- The first solar flare is peaked at 1841 UTC on 10/25/21, four minutes after the Pollux/Mercury conjunction peak. It is located at an AR that was about 90 degrees behind Mercury's position. This seems to relate to not just one X-ray flare but a series of them. Pollux is only 38ly away from our Sun and is almost twice the mass of our Sun.
- Another concentration of flares happened after the conjunction of Mercury with the star ȝ Cnc (Chi Cancri) 59.4ly away and +0.7 degree on 10/27/21 at 0215 UTC. This star is just slightly bigger than our Sun.

Without a doubt, these statistics fascinated me and my interest. After all, I had been researching the Sun and Mercury—a lot.

As I stated earlier, I use real-time data and sometimes from my computer screen print an image that will not be sharp to view. The key point is displayed, however. There was a correlation with the Mercury/star conjunctions and a series of solar flares. Also, this is not the first time I have noticed solar flares happen quite often in groups. That is not randomly distributed. More detailed information on this phenomena can be found at spaceweatherlive.com in the archive solar flares information.

Solar Tsunamis on the Hot Planet

All this incredible solar activity with Mercury reminds me of a term called "solar tsunamis," aka Moreton wave. It's a sign of a large-scale solar coronal shock wave caused by a solar flare. These waves can cover the whole Sun. The phenomenon reminded me of the MEGA-waves scene from the movie *Interstellar*. One character notes the tsunami is mountains when her colleague corrects her and exclaims, "They're waves!" as a life-threatening gravity-induced tsunami moves toward the astronauts and their spacecraft.

A solar tsunami is an invisible magnetic wave from a solar flare. And yes, the big-screen happening was make-believe, but in real life, there could be a chain-reaction situation, just like this. When the Sun has active regions, it also can have solar flares, which cause tsunamis that travel all over the Sun, affecting other active areas on the Sun. The traveling tsunamis, which happen under the right circumstances, could create a new flare and solar tsunami, which then starts the whole sequence over again somewhere on the Sun that has an active region. (1)

Interestingly, this solar-tsunami action is comparable to a nuclear chain reaction that occurs in our nuclear power stations on Earth. Add to this the fact that a coronal hole will stop the progress of a solar tsunami—at least its visible plasma, and probably the magnetic fields because the coronal holes are high, Sun-exiting, solar-wind areas. This would cause the tsunami to deviate from a circle and develop a wave front that isn't curved like the edge of a circle.

Eye-Catching X-Ray Spikes and Mercury

X-ray spikes can happen after a star conjunction with Mercury. How? How does it do that? The following data shows the opposite may be true at times:

Images of Mercury in relation to Earth show in the first graphic. The second graphic and data are derived from the GOES satellite X-ray detection system, which is near Earth. (The GOES-R series is run by the National Oceanic and Atmospheric Administration and NASA, and provides atmospheric measurements of Earth's western hemisphere and space weather monitoring.) This visual information pins down the time of the solar flare. The X-ray flare detections usually are related to a solar flare. The third graphic is the AstroGrav data I used to pin down the time of the conjunction. The fourth graphic shows where on the Sun the flare happened and is not usually found in my notes.

Images from theplanetstoday.com with permission from Hayling Graphics Limited

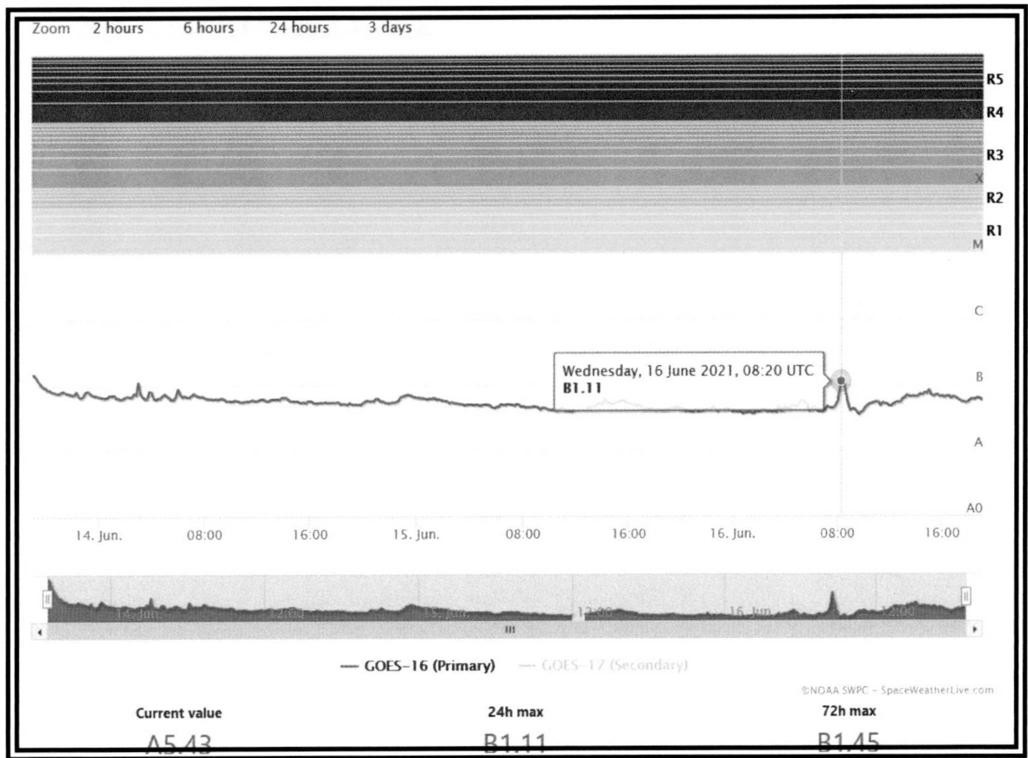

National Oceanic & Atmospheric Administration (NOAA) – spaceweatherlive

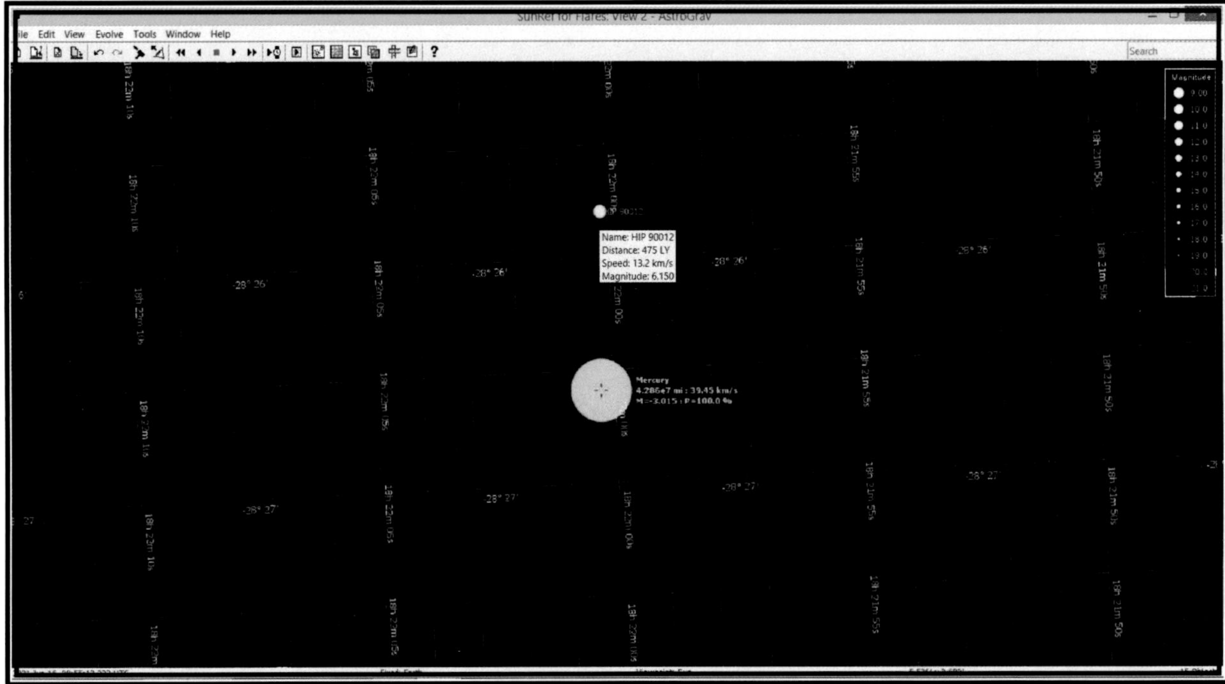

AstroGrav software generated

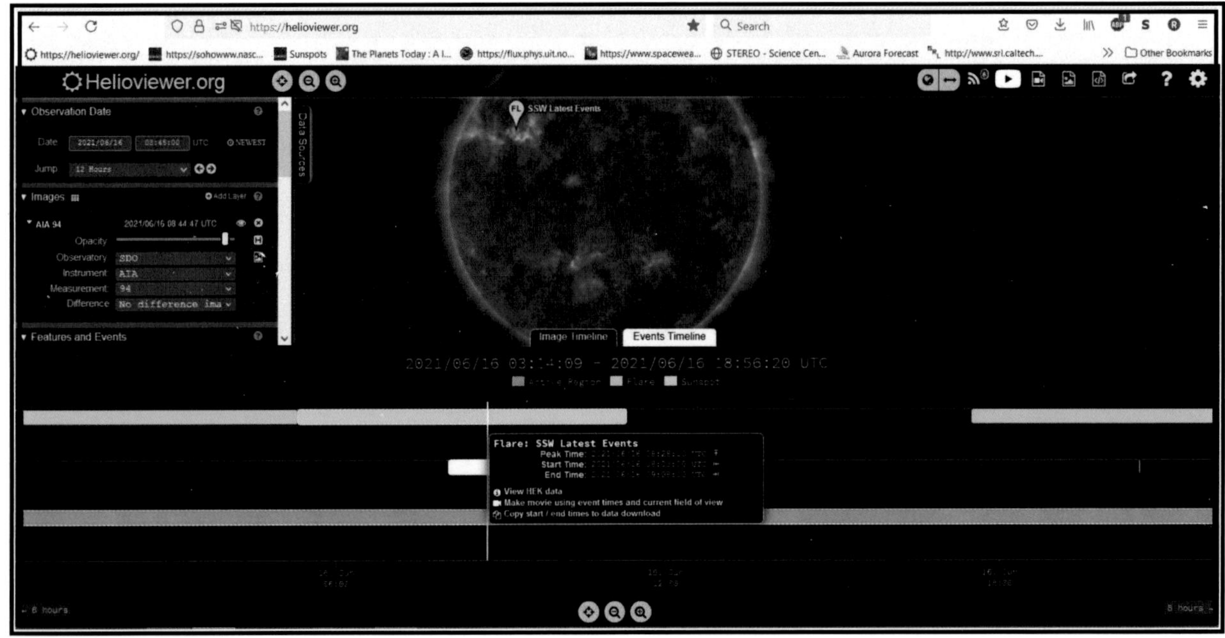

Helioviewer – Open Source Data

The previous 4 graphics are related to a conjunction of Mercury with the star HIP 90012. The X-ray spike is during a very calm period on the Sun. It was caused by Mercury being in conjunction with HIP 90012 475ly away on 06/16/21 at 0855 UTC. This spike is 45 minutes before the conjunction, minus 8.5 minutes for X-ray travel to Earth, which means the Sun reacted 36.5 minutes before the conjunction. This conjunction also produced a solar flare on 06/16/21 at 0828 UTC. This flare started at 0803 UTC and ended at 0908 UTC on 6/16/21.

Magnetic Mercury, One of Its Kind

Mercury is distinctive and matchless from the other planets. The most important difference is, it is made of mostly iron. Iron affects magnetic lines differently from the other elements. Also, Mercury is closest to our Sun and of course orbits our Sun the fastest. That means the exposures to the Shaylik from stars beyond our Solar System will be the shortest in time of all the planets. Also, when Mercury is between Earth and the Sun and there is a CME headed toward Earth, Mercury will most likely get hit by the CME well before Earth will and will get statically charged. The discharge of this static charge could be what causes the first in a series of solar flares. While we know Mercury is the smallest planet in our Solar System, it's only slightly larger than Earth's Moon.

Bits and Pieces of Magnetic Matters

- *Like a Magnet:* The fact that Mercury is 70 percent iron, which I suppose is its core, is interesting. Iron is the element that Shaylik seems to like the best. For Shaylik, it acts like the beginnings of a black hole (this is where gravity has become so strong that nothing around it can escape, not even light). The element iron is a magnetically malleable substance. That is, it can at least partially cooperate with the magnetism of whatever magnetic fields are around. The electrons in iron can move to accept any chirality at any frequency. So instead of my small neodymium magnet being repulsed no matter what position it is in, it's accepted into this metal like it is a magnetic black hole. All the other magnetic forces around me are pushing my magnet toward the iron.
- *Magnetic Bubbles:* Did you know magnetic bubbles emanating all around me and the rest of the cosmos are pushing against each other? The walls of the building, the table, the chairs, and the magnet are all pushing on other magnetic bubbles in the area. Magnetic bubbles are said to exert forces infinitely in all directions. But, alas, they cannot be infinitely large because there are always other magnetic bubbles to push them back. In fact, the biggest magnetic bubbles we know of are out where our farthest space probes have gone beyond Pluto. "The sausage-shaped bubbles are gigantic, measuring about 100 million miles (161 million km) across. And there are a lot of them." So because, at least in our Solar System, there are magnetic bubbles everywhere, according to stacks of research, there's magnetic pressure everywhere. All are repulsive and are jostling for their place in space. And they can get fairly small. "The bubble logic would use nanotechnology and has been demonstrated to have access times of 7ms, which is faster than the 10ms access times that present hard drives have." (2)
- *Magnetic Pressure:* We have an example of how this works here on Earth. It is called "atmospheric pressure," and it is 14 pounds of pressure per square inch at sea level. So if the magnet I am holding near the iron is presenting one inch of surface area and the iron is a place that is zero magnetic pressure, the magnet will go toward the iron, like it is attracted to it, while actually it is being pushed. And it will go toward the iron with a force of 14 pounds! This is just another outside-in

thought in progress. But I'm certain that the magnetic environment of planet Mercury is unique because of the preponderance of iron and its position close to the Sun. I am getting hints from the data I am accumulating.

X-Ray Spikes, Mercury, and Shaylik

Often, not unlike the timing of planet alignments, there are X-ray spikes that are timed well with the conjunction of Mercury. I need to keep in mind it's pinpointing that this is indicative of a solar flare presumably somewhere on the Sun. The spot on the Sun that is chosen by Mercury's energy dump may be based on the chirality of the energy finding a compatible chiral spot on the Sun.

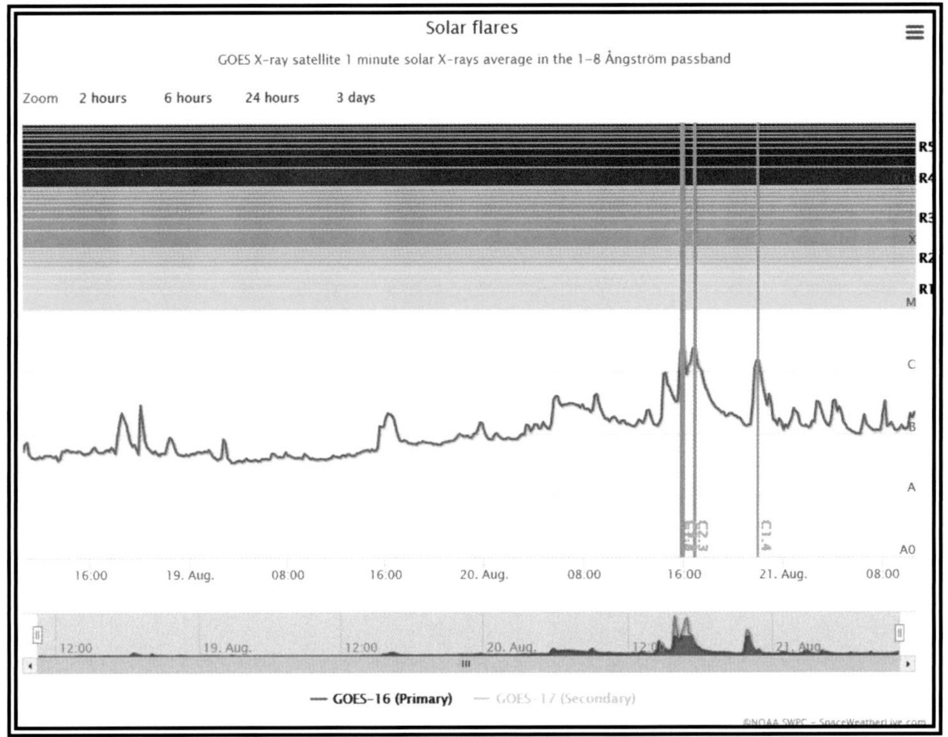

National Oceanic & Atmospheric Administration (NOAA) – spaceweatherlive

On 08/20/21, there were four stars that were in a one-quarter-degree conjunction with Mercury, and the timing correlated with four X-ray spikes detected by a satellite near Earth. The location of the presumed solar flares was not determined by our solar scientist for one of the four flares, so they showed the location as 0,0, which is their default location for unknowns. They are looking for a visual clue on the Sun, and sometimes they do not find one. The 0,0 location is directly in the middle of our Sun at the time. Two flares were located on the trailing limb of the Sun, almost directly across from Mercury's location. I suspect flares may happen on the Sun that are located in a direction opposite the location where the planet involved is. And furthermore, it's surprising to see so many unknown locations happen.

Unknown is just that. The source could be from anywhere on or nearby the Sun where the satellite is looking for X-rays. So this makes it possible the X-rays did not originate on the Sun. They could have come from the Shaylik between the Sun and Mercury, or from Mercury itself. I find this a very interesting thought since I have observed that X-rays are the only light that seems to interact with Shaylik. If there are X-rays coming from the excited Shaylik between Mercury and the Sun, I would not be surprised if it is the shape of a sigmoid, (S), nearest the Sun, with the curves of the sigmoid in line with the Sun's rotation. So pictures of this could be everything from a straight line to a wiggly line to a full sigmoid, depending on the camera's perspective.

It's unlikely that X-ray light will be detected from the other side of the Sun because it is light and will not curve around the Sun. So from now on, I will be looking for hints/data that tell me if it is Mercury or Mercury's Shaylik. This will of course be affected by where Mercury is and if it is shadowed by the Sun from our view here on Earth. If I see more than usual unknowns while Mercury is shadowed, then I know I am on the wrong track. However, if the opposite is true and there are fewer unknowns when Mercury is shadowed, I may be onto something. This gives you an idea of how I think through these situations. Logic is my key tool at this point.

By the way, from our perspective on Earth, Mercury does not get physically behind our Sun very often. The orbital period for the two planets being opposite the Sun is about 124 days. On most of those occasions, Mercury is not blocked by the Sun visually. This means it will be a while before I can get any definitive data relating to possible Mercury-related X-rays.

Gazing at Four Stars of Mercury

Enter an X-ray response to a small cluster of four stars that Mercury came in conjunction with and the four X-ray spikes shown with yellow C levels and spike tips on 08/20/21 are what happened. The stars' conjunctions lasted from 08/20/21 at 1643 UTC to 1827 UTC, ending with the closest star, HIP 68038 at 189ly. All the stars were within 0.2 degree of mercury at the time of conjunction. The spikes' peaks are at 1540, 1550, 1650, and 2150 UTC. So the four X-ray spikes started one hour and three minutes before and ended three hours and 23 minutes after the conjunctions. Mercury is about -100 degrees behind Earth at this time. Venus was in conjunction with the star Acrab, 530ly away at 1740 UTC on 08/20/21, and the star B2 Sco at 904ly away. Venus was also behind Earth at about 90 degrees. Earth wasn't in conjunction during this period, nor was Mars.

FOUR STAR CONJUNCTIONS IN ORDER

- HIP 67968 08/20/21 1643 UTC
- HIP 68001 08/20/21 1716 UTC
- HIP 68024 08/20/21 1750 UTC
- HIP 68038 08/20/21 1827 UTC

So the thing is, Mercury may or may not be the definitive missing puzzle in the four-star picture. The information I've shared with you is the best example as I live in the moment—and see what's happening to our Sun. Time will tell.

The fact that there are four stars' conjunctions with Mercury and four solar flares at about the same time is something I can't ignore, nor should you. Plus, Venus had two conjunctions at the right flare times and must be considered also.

Images from theplanetstoday.com with permission from Hayling Graphics Limited

OH! There's another possibility I've pondered for detecting X-rays from excited Shaylik. Just like Mercury, as a satellite of our Sun, the GOES spacecraft is in geosynchronous orbit as a satellite of Earth and can get statically charged. This satellite has the detector of the solar flare–caused X-ray spikes we see on the spaceweatherlive.com site. The geosynchronous orbit means that it is opposite Earth's equator and is about twice as far away from Earth as our Moon. This also means this satellite will be hit by some of the same CME energy that hits our Earth. There could be times when the static charge on this satellite is sufficient to cause a static discharge between it and Earth or our Sun. If the static charge happens when Earth and the satellite are in the Shaylik between a nearby star and Earth or the Sun, there could be a detectable emission of X-rays that appears as a solar-flare spike but of course is not. This then would be designated as an unknown-location solar flare that has about the same timing as the Shaylik conjunction. I'll be watching for this possibility also.

The final word is, the data I've shared with you is exciting! It's encouraging because the idea the Mercury/star conjunctions may be what causes solar flares to solar winds. That means it is like an earthquake or hurricane forecast tool—a heads-up to prepare before it happens. Meanwhile, the jury is still out. More scientific data is required to come to a reasonable conclusion about this idea and Mercury's role in solar chain reactions.

While solar flares and solar winds are fascinating, it doesn't stop there. The Sun is prone to creating the perfect storm that can—and does—affect Earth, as you'll discover in the next chapter. No telescope needed.

NEW THOUGHTS, OLD UNIVERSE CLUES

- ✓ A conjunction of the Sun, Mercury, and a nearby named star can correlate with the timing of a solar flare somewhere on our Sun.
- ✓ A year on Mercury takes 88 Earth days; a day on the surface of Mercury lasts 176 Earth days.
- ✓ Mercury has a diameter of 4,879 km, making it the smallest planet.
- ✓ Venus is the second planet from the Sun—it's the hottest planet in the Solar System, not Mercury. Why . . . ?
- ✓ Mercury doesn't have any atmosphere that can trap the heat like Venus can and does.

CHAPTER 11

Solar Superstorms

And the Earth, turning upon itself, moves round the Sun.
—Galileo Galilei

As a boy in first grade was the first time I was introduced to the Sun. Our class learned that the Sun is the star at the center of the Solar System. As time passed, I was also taught it's a sphere of hot plasma, heated to incandescence by nuclear fusion reactions in its core, radiating the energy mainly as light and infrared radiation. But as time passed, I discovered that's not all . . .

The idea of the Sun and superstorms is reality too. Solar tornadoes are not a new thought for me. Why? Because I see the Shaylik that is coming into each Moon, planet, and star as a twisting double-helix. The effect of this on an atmosphere would be to at least temporarily create a tornado—a solar tornado. So thousands of them around our Sun is actually welcome supportive news.

The Sun Is Hot Stuff

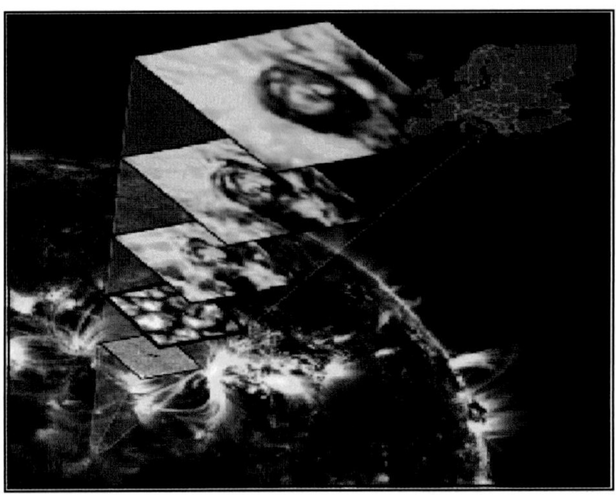

The full article attached to this photo can be found at the link provided.

The discussion is centered on the cause of the Sun's extra-hot corona. The authors speculate that there are at least 11,000 vortices on the Sun's surface as a minimum. They do not say they are magnetic, just that they are there with diameters up to 1,500 km. This structure, the vortex or tornado, is what I expect is the shape of the double-helix structure of Shaylik that is coming into a mass like Earth or our Sun and shepherds the atoms. They detected this phenomenon from the atom's movement within the vortices, which is the visible structure involved.

There is an admission that they do not know what is causing the coronal heating and these vortices could be the cause. They also speculate that the vortices are generated by energy inside the Sun.

My speculation is that there's eddy-current heating going on in the corona that is somehow more sensitive to the eddy-current effects of the magnetic double-helix structure. Perhaps it is a combination of magnetic fields from the Sun and the Shaylik coming into it that is most dense in the Sun's corona. Time will tell which is correct.

(For more information: https://arstechnica.com/science/2012/06/why-is-the-suns-corona-so-hot-two-words-solar-tornadoes/)

Scientists and you, too, may have wondered, why is the Sun's corona so hot? The answer may be solar tornadoes. These volcano-ish vortices or twisters in the Sun's atmosphere may be from twisted magnetic fields—and create unimaginable heat.

The Ars Technica research clearly points out there are thousands of magnetic vortices coming out of our Sun. Are they coming out or going in?

This could be evidence of the Shaylik I've proposed coming from the surrounding nearby stars, while turning in a double-helix shape, a shape that tapers down to our Sun's corona due to competing double-helix vortices trying to fit in a smaller and smaller space. The article states, "Each vortex was aligned along a single axis." These words suggest a straight line, and it is one of the features of Shaylik between masses. (1)

Extra Matter on Solar Tornadoes

The number of solar vortices is significant to the understanding of how two-dimensional Shaylik becomes concentrated and more dense as it approaches a planet or star. The idea that the observed solar vortices are caused by energy from the Sun is still open to reasons for why it may be correct. There's a possibility that the idea that Shaylik forms a straight line between masses and is a rotating magnetic structure could lead to new discoveries.

It's estimated that there are about 2,000 stars within 50 light-years of our Sun and 113,000 within 100 light-years. The majority of these stars are smaller than our Sun and known as brown dwarfs or failed stars. So it's quite possible that there are 11,000 stars as massive as, or more massive than, our Sun, within 100ly, that are causing the solar tornadoes on our Sun.

Large, Earthbound tornadoes are super-impressive weather events. Those dark, giant funnel clouds inspire much awe and fear. But they're nothing compared to so-called solar tornadoes' solar gases swirling in monstrous formations deep within the Sun's atmosphere, measuring the width of several Earths and spinning at up to 190,000 miles per hour.

Scientists have known about solar tornadoes for some time. But just recently, they've captured a massive one on camera for the first time. Unlike earthly tornadoes, which consist of powerful swirling winds, solar tornadoes are magnetic events involving superheated gases being sucked up from the Sun toward its upper atmosphere.

Now, earthly tornadoes are noteworthy for the material damage they can cause. While solar tornadoes are no threat to trailer parks, they may be implicated in other sorts of mayhem.

Scientists believe that solar tornadoes often occur in conjunction with a coronal mass ejection when a part of the Sun's atmosphere breaks off and jets through space. These ejections cause solar storms that can make satellites go on the fritz and even affect Earth's electrical grid. So the more scientists understand about what causes solar tornadoes and how they work, the better able they may be to predict solar storms that can damage man-made technology. (3)

Sensational Solar Light Show

The Sun has fascinating features: solar flares, coronal mass ejections, active regions, and solar wind. Solar flares happen in areas that have active regions and will out of the blue appear as bright spots. According to NASA experts, these flares can cause high-energy particles to emit from the Sun, which can wreak havoc on astronauts and cause damage to satellites orbiting Earth. Solar flares emit bursts of electromagnetic radiation, including high-energy X-rays and gamma rays. The energy released by one flare can be more powerful than a million nuclear bombs. These particles can also damage electronic components and affect radio signals.

Then coronal mass ejection—solar flares are followed by a large ejection of plasma from the surface of the Sun. These are called "CMEs."

Next up are solar winds. The Sun is so powerful and energetic that it actually creates a type of wind that travels through the Solar System. The wind is called the "solar wind." The power of the solar wind varies depending on the activity of the surface of the Sun. Earth is protected from solar wind by its strong magnetic field.

Last but not least are the beautiful Northern Lights, aka aurora borealis. These lights seen in the dark and clear night are caused by the solar wind hitting Earth's atmosphere. The vibrant, moving colorful lights—green, purple, and white—are usually seen in the northern hemisphere. Northern Lights are often viewed at a latitude of 66 to 69 degrees north, such as Alaska and Canada, in November to March. The auroras happen in a band known as the auroral zone, which happens when solar activity is high. It's best to check out the kp index (a measure of electromagnetic activity in the atmosphere). If a reading is of two or higher, you are more likely to view a solar light show.

So when did I get interested in solar rain? I got interested about a year ago, when I first saw the solar rain video. We had already decided that Shaylik was a forever-flowing, like a river, phenomenon, and this was the first evidence of that flow. This video shows the visible plasma within the invisible Shaylik coronal loops coming from above, out of seemingly nothing! This we see as Shaylik changing from a two-dimensional form to a three-dimensional plasma at the top of the phenomenon. All motion from there is down toward the Sun.

NASA video clip **(FYI:** https://www.youtube.com/watch?v=1ZxrVlegJ28&ab_channel=NASAVideo) (2)

Raindrops Keep Falling on the Sun

Wow! Rain loops on the Sun are an amazing phenomenon. It looks like a never-ending upside-down volcanic eruption with vibrant red and yellow, hot, flowing magma. Certainly NASA knows this, I know it—and now you do too. (2)

It's obvious that something is coming into the top of these magnetic loops of plasma on our Sun. What I've learned about this phenomenon is that some scientists think magnetic lines are coming from elsewhere on the Sun and coming into this structure's top. OK, this may be true.

I believe it also may be spot-on and more likely that this coronal rain is due to the invisible Shaylik coming from a large nearby star or planet and interacting with Sun's plasma (a state of matter), which makes its effects visible. This then would just be a normal day on the Sun, where forces from outside the Sun are causing noticeable changes to the Sun's corona or visible surface.

So I see in both of these examples that there is at least two-dimensional Shaylik coming into our Sun and changing to three-dimensional mass as a plasma or mixing with the plasma that is already there and moving it along with the remaining two-dimensional Shaylik. This is support for the concept that Shaylik is flowing between at least Solar System masses.

While research will continue to find out more about solar tornadoes, there's no such thing on our planets. Some folks believe the Sun could just fade out one day and that would be the end of solar tornadoes—and the end of Earth.

Now that you've gotten a glimpse of the Sun and its related cause and effect of solar superstorms, it's time to get grounded and come back to Earth. Read on—you may be awestruck about more happenings.

NEW THOUGHTS, OLD UNIVERSE CLUES

- ✓ Solar vortices, also known as tornadoes, happen around the Sun—and are linked to Shaylik.
- ✓ Solar rain flow of visible plasma within invisible Shaylik are coronal loops . . .
- ✓ And are the force behind solar flares.
- ✓ Evidence shows solar tornadoes can happen in conjunction with a CME when a part of the Sun's atmosphere breaks off and zooms through space.
- ✓ The Sun plays a role in an incredible light show, including solar flares to Northern Lights.

PART 5

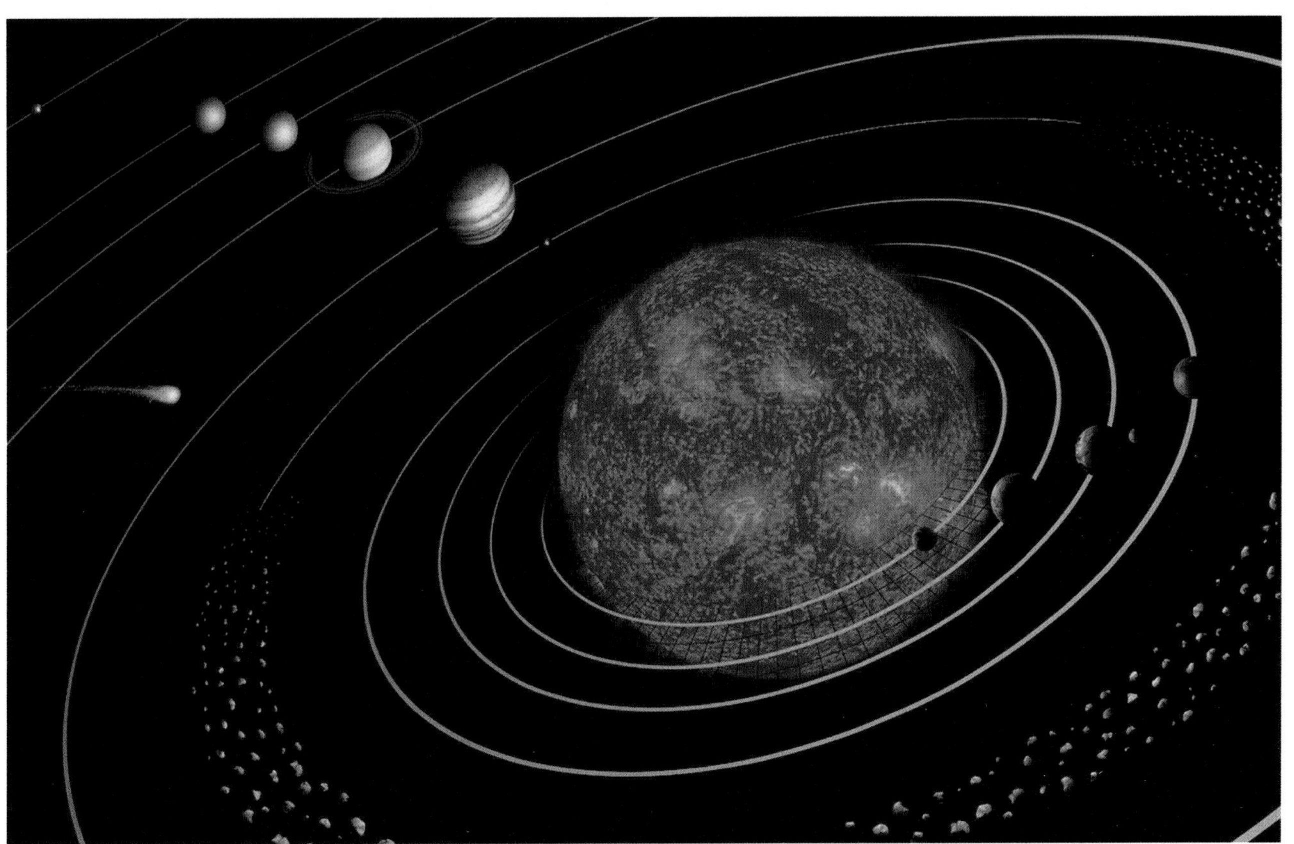

MORE MATTERS OF SHAYLIK

CHAPTER 12

Electrifying the Earth

There is peace even in the storm.
—Vincent van Gogh, *The Letters of Vincent van Gogh*

Imagine this scenario: A man witnessed an eventful storm. He viewed a very large solar flare "megaflare" and was the first to realize the link between activity on the Sun and geomagnetic disturbances on Earth. The storm was coined the "Carrington Event" after this man, a British astronomer named Richard Carrington. As the story goes, in the morning, he saw a sudden flash of intense white light from the areas of sunspots. Hours later, the night sky in North America lit up like daylight. Spikes of electricity surged in the world's telegraph system (used for transmitting messages from a distance along a wire) and nobody could communicate—think of a mega internet outage in the twenty-first century and how it affects life as we know it. Today, such a happening could damage satellites; disable communications by phone, radio, and TV; and cause blackouts (all the things that create heart-pounding drama in a sci-fi film).

Back in the 1800s, scientists didn't understand what caused auroras and the electrical and magnetic disturbances. Solar storms, like this Carrington Event, can blast out huge clouds of electrified gas and dust at up to two million miles an hour. If high-energy blasts of particles reach Earth, they can distort and disrupt Earth's magnetic field. (1) (2)

In present-day, I've discovered this historical phenomenon through logic and luck early on in my research. I used the Planets Today online software to first look back in time at the moment of the Carrington Event. Then I recalled where Earth and Mars were and I moved the display forward in time until I saw the same locations for both planets.

The first time I did this study, I actually missed the time it happened in 1938 and found it in 2017. Later, I realized this is a harmonically related phenomenon and simply went back in time half the difference between 2017 and 1859, in early September.

Segments 1 and 2 ("The Carrington Event")

The Carrington Event is fascinating. It consists of solar flares and coronal mass ejections. (Refer to chapter 11 to reboot your brain to the reasons they twist and short-circuit themselves.) Solar flares are like huge lightning bolts or electrical short circuits. What's more, when these flares happen within the Sun's photosphere, usually associated with an active region, some of the Sun's surface structure can be launched off and away from our Sun. A portion of the plasma could fail in leaving our Sun and fall back down to make a huge splash and create a solar tsunami.

Occasionally, CMEs strike Earth and can cause major changes to the magnetic field, which can ruin our electronic-age infrastructure by overloading it. (Think of your home when it blows a fuse. The result? A house power outage.) Well, when a CME hits our planet, it's the same but on a much larger scale. Widespread blackouts to power grids occur; and travel by air or on the road can be disrupted.

The direction the CMEs leave the Sun is very random. They can be blown straight up or be launched sideways, and everything in between. The proof is in the history set at the time of the Carrington Event. This event was when a part of our Sun hit Earth. This caused disruption of electronic communications over most of our planet. Only a telegraph was used back in the nineteenth century, and it caused telegraph pylons to spark and gave shocks to operators, as history tells it. And note, if this type of solar event were to happen today, communication systems we have now would be severely damaged, causing a great upheaval in the way we live.

Images from theplanetstoday.com with permission from Hayling Graphics Limited

The graphic is from the online program the Planets Today, with it set at the time of the Carrington Event. This event was when a part of our Sun was blown off the Sun and hit Earth.

Take a closer look and notice where Earth, Mercury, and Mars are at this time. Now compare the Earth/Mercury/Mars locations with the next graphic, dated 09/05/2017. Mars is red, and Earth is blue.

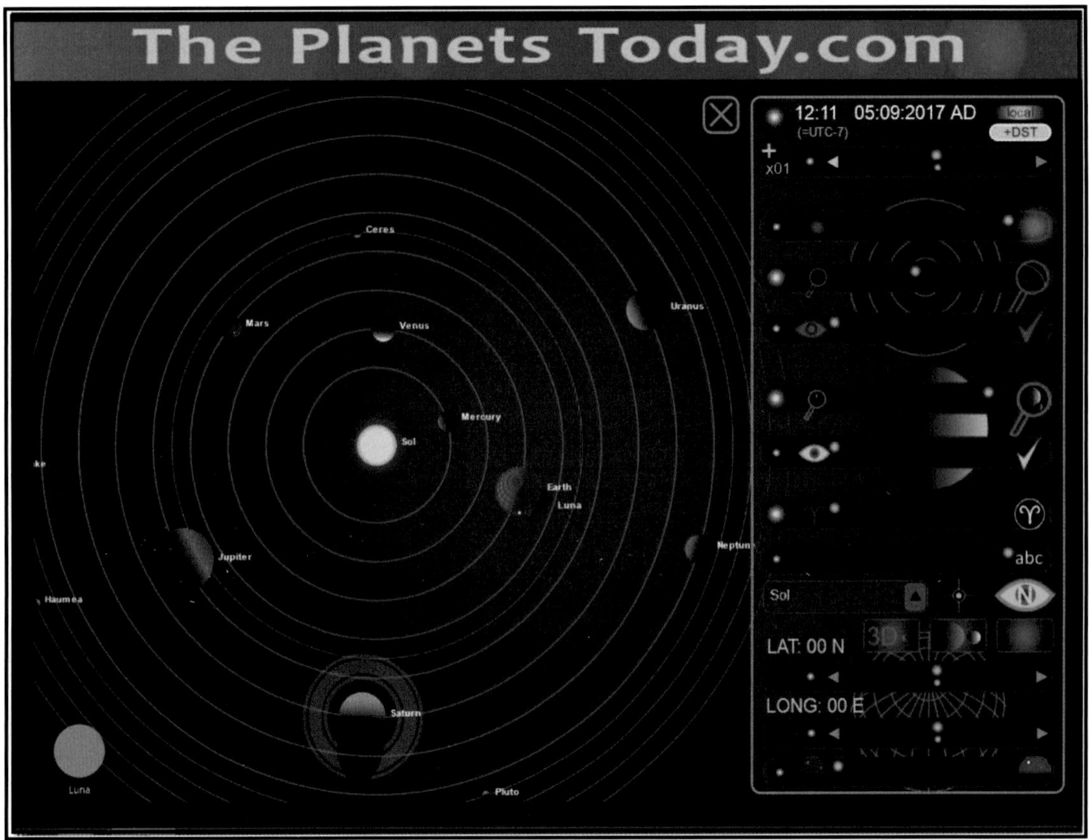

Images from theplanetstoday.com with permission from Hayling Graphics Limited

If you explore the past happenings, when Earth and Mars appear to be in the same location and upon more detailed investigation, they are nearly exactly the same except the day of the year is three days later in the 2017 graphic. On both occasions, Earth was hit by a CME from our Sun. On both occasions, Earth was in conjunction with the star TYC 583-236-1, which is 33ly away, and the star Lambda Aquarii, which is 390ly away. Interestingly, four objects in our Solar System were in line with each other: Our Sun, Earth, TYC 583-236-1, and Lambda Aquarii on both occasions also coincided with a CME striking Earth and caused problems with electronics.

Also, Mars is noted because a CME on the Sun requires what is called an "active region" in order to experience a solar flare in it that can launch a CME toward Earth. I suspect that Mars was in conjunction with a star or stars that created an active region on the other side of the Sun about 12 days before it came around to the Earth side and participated in the launching of a CME.

Carrington CME—09/02/1859 1300 hours local time = 2000 UTC

- Earth/Lambda Aquarii 306ly -0.4° ecliptic conjunction 09/02/1859 at 1452 local time.
- Mars/HIP 45699 128ly +2° ecliptic conjunction 08/27/1859 at 1321 local time.
- Jupiter/HIP 32333 349ly -0.3° ecliptic conjunction 08/21/1859 at 1652 local time.
- Saturn/Mars/HIP 46232 272ly conjunction 09/01/1859 at 0052 local time.
- Mercury/30 PSC, 415ly -0.25° conjunction 09/02/1859 at 0932 local time.
- Mercury/33 PSC, 129ly -0.3° conjunction 09/02/1859 at 1402 local time.

Carrington 2—09/05/2017 0330 local time CME = 1030 UTC

- Earth/Lambda Aquarii 306ly -0.4° ecliptic conjunction 09/04/2017 at 0424 local time.
- Mars/HIP 45699 128ly +2° ecliptic conjunction 08/24/2017 at 1421 local time.
- Earth/Neptune -1.8° conjunction 09/05/2017 0501 local time.
- Jupiter/HIP 67271 512ly +2° ecliptic conjunction 08/25/2017 at 2101 local time.
- Saturn/HIP 86819 251ly 0.1° ecliptic conjunction 08/25/2017 at 0401 local time.
- Mercury/HIP 3540, 134ly +0.1° conjunction 09/03/2017 at 1846 local time.

So I see these two events are similar in terms of the location of Earth and Mars, and both were when a CME from our Sun hit Earth. Also, I see that Mercury is in a slightly different location. The other planets are not even close to a similar location. It is also interesting that Mercury was in conjunction with a large nearby star in the 1859 Carrington event, and in the 2017 event, it was in conjunction with a small nearby star. Couple this with the fact the Carrington Event had a much-greater effect on Earth than the 2017 event. It may be that Mercury was involved with both these events.

Because of this thought and other data I'm gathering on Mercury/conjunction/flare events, Mercury has my full attention. In comparison with the other planets, Mercury has the most events of a Mercury/nearby-star conjunction timed with a flare on our Sun. Also, there are many times when those conjunctions are not associated with a location on our Sun. When there is an X-ray flare associated with an unknown location, the solar scientists show the location as 0,0, which is the center of our Sun at that moment.

Preliminary data has already borne fruit. Mercury is definitely the planet that's most likely to have a flare on the Sun associated with a nearby-star conjunction. Also, after observing only one orbit of Mercury, it appears that when Mercury is more than 150 degrees ahead or behind Earth in its orbit, it stops having conjunctions correlate with solar flares. This suggests that some of the unknown-location solar flares are because Mercury is on the other side of the Sun from Earth. Could it be that the source of the X-rays is coming from Mercury or more likely the Shaylik between Mercury and the Sun?

I say "more likely" because I have long suspected that two-dimensional Shaylik is superconductive, which simply means it conducts electricity with zero resistance. And I also know that lightning strikes here on Earth produce X-rays. Put this duo together with the thought that all planets as they are going around the Sun are being statically charged. I've seen this trio set up as an obvious possibility for many years now.

Our cars and airplanes get statically charged when they pass through Earth's atmosphere. So why wouldn't the planets get charged while passing through the Sun's solar winds on their way around the Sun? The greatest charging would be when Earth passed through a CME from our Sun.

Lightning strikes (imagine vibrant white streaks of fireworks in the dark sky during a storm) are visible on Earth because the atmosphere is lit up from the energy. A similar bolt of energy from Mercury to the Sun may be undetectable in all but X-ray light. This could explain the unknown location of flares on the Sun and why these flare readings diminish as Mercury is on the other side of our Sun from Earth. Either the Sun blocks our view of the X-lightning from Earth or we are not seeing the X-lightning from the Shaylik because Earth's angle to the Shaylik line is too shallow an angle, making the signal very weak. Another point is that Mercury, being the closest planet to our Sun and made of mostly iron, is the most likely to get statically charged by the most intense exposure to the solar winds. One more thing: Many constituents of solar winds are also in our atmosphere.

So my preliminary thoughts on the two CME events are a work in progress. Going back to the event, I can analyze why the actions happened. First, Mars is in conjunction with HIP 45699 128ly away while it is on the other side of the Sun from Earth. This causes an active region to develop on the opposite side of the Sun from Earth. Then, after about 12 days, the active region is now directly across from Earth and now in the Shaylik between Earth and the Sun. Now, simultaneously, Mercury has a good static charge because no conjunctions have been able to discharge it. Mercury comes into conjunction with a nearby star, releasing the static energy down the Shaylik to the Sun. X-rays are emitted by the X-lightning, and this side of the Sun is temporarily charged by static jolt. This then causes the Shaylik link between Earth and the Sun to blow a part of the active region off the Sun toward Earth via a solar flare explosion. Hours or days later, Earth suffers a magnetic jolt from the CME passing by. These are seeds of thoughts, which may spawn the end result of why Shaylik is part of the event. I am enthusiastic like astronauts on the edge of a new discovery.

Incidentally, it's well known that all of our planets' orbital periods are harmonically related. This just means that this alignment of planets and stars will happen time after time. Then again, stars involved will move slightly each time, but the planets should be repeating locations almost exactly. The repeat period related to the Carrington event is 79 years and two days. So around 09/05/1938, we had a conjunction like the Carrington Event that was apparently uneventful. And in the future, this combination of stars and planets will happen on approximately 09/07/2096. While I, and likely you too, won't be here in about 80 years, our great-grandchildren will be able to experience the event—and other happenings too.

Segment 3 ("CME Hits Earth")

On the morning of August 4, 1972, a CME from our Sun hit Earth. The result: It was the cause of explosions of at least 12 magnetically triggered sea mines in a Vietnam harbor. If I assume the hour this happened was about 7:00 a.m., and the reported CME took 14.6 hours to reach Earth from the Sun, then this CME left the sun on August 3 at 0824 UTC. This CME was apparently related to the third and biggest of four solar flares happening between August 2 and August 4.

(FYI: https://www.colorado.edu/today/2018/11/12/1972-solar-storm-triggered-vietnam-war-mystery)

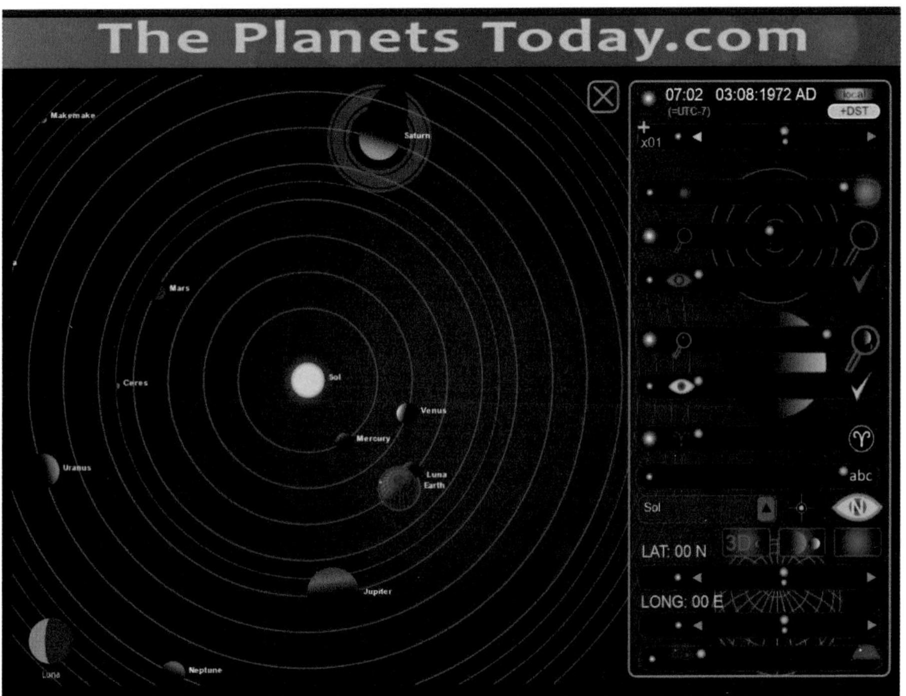

Images from theplanetstoday.com with permission from Hayling Graphics Limited

Note the positions of Earth and Mercury are nearly the same angle. Furthermore, on 08/02/1972 at 2248 hours UTC, Earth was in conjunction with the star 19 Cap. This star is 397ly away from our Sun, and the conjunction was within 0.5 degree. This combination could have caused the active region on the Sun to become more dynamic. Then, 13.5 hours later, on 08/03/1972 at 1219 UTC, Mercury was in conjunction with HIP 100311 745ly away and 0.3 degree apart. This lineup could have caused the solar flare that hit Earth 14.6 hours later on 08/04/1972 at 0255 hours UTC or 0955 hours Vietnam time. (3)

Supporting all this was Jupiter, which was on the same side of the Sun as Earth. Jupiter was, on 08/04/1972 at 2232 hours UTC, in conjunction with the star HIP 90793 505ly away. Both were on the ecliptic and 0.2 degree apart. And earlier, Jupiter was in line with the star HIP 90713 2560ly away and only 0.1 degree apart on the ecliptic. This was on 08/02/1972 at 1432 UTC. Also, on 08/07/1972, Earth and Mercury were in conjunction but Mercury was -7.0 degrees below the ecliptic and Earth. (4) (5)

Past research suggests the same active area was long-lived. It persisted through the five solar rotation cycles. The August 4 flare that triggered the extreme solar particle event (SPE) around an intense geomagnetic storm on Earth was among the largest documented by science. It's also been reported the arrival time of the associated (CME) and its coronal cloud, 14.6 hours, remains the recorded shortest duration as of November 2018, indicating an exceptionally fast and typically geoeffective event (the normal transit time is merely two to three days). And that's not all . . .

Segment 4 ("Summer Solar Superstorm")

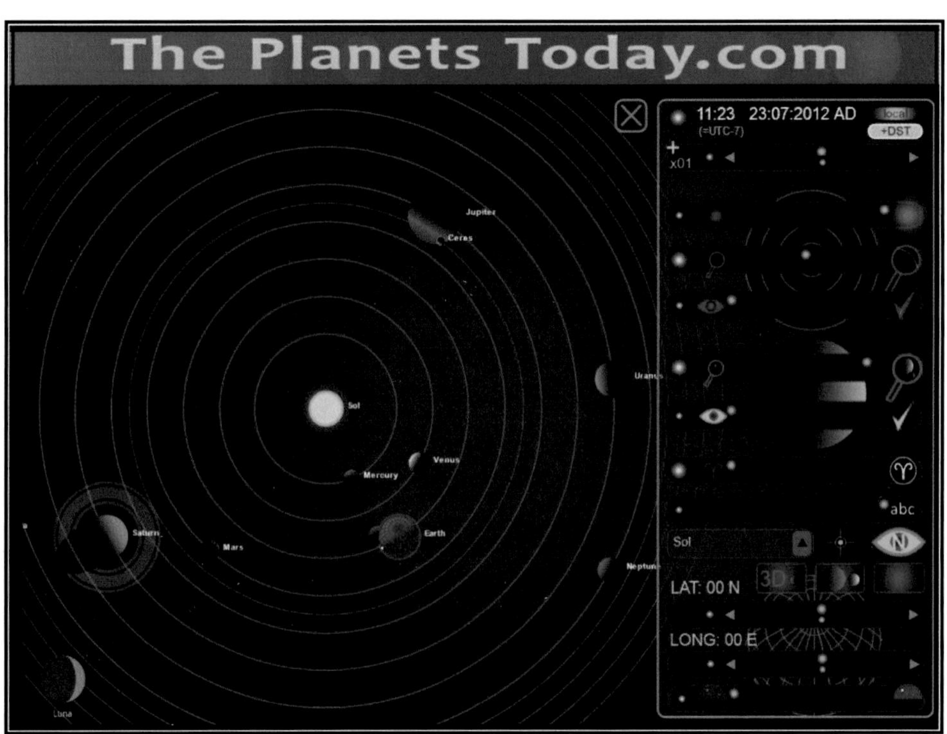

Images from theplanetstoday.com with permission from Hayling Graphics Limited

(FYI: https://www.spacedaily.com/reports/Near_Miss_The_Solar_Superstorm_of_July_2012_999.html)

There was another noteworthy solar superstorm on 07/23/2012. Note the positions of Earth and Mercury are nearly the same angle. The time is adjusted to UTC. Mercury was in conjunction with HIP 96124, 90.7ly away and 0.002 degree apart, on 07/23/2012 at 2107 UTC. A C2.0-level flare happened at 1127 UTC on the same day, and a flare, at B7.9 level, happened at 2225 UTC on the same day. So the largest flare happened one hour and 18 minutes after the moment of conjunction. (6)

So like an excited bloodhound on a mission, I've tracked quite a few events involving both Earth and Mercury, which were in conjunction with various nearby stars, and at least one active region on the

Earth-facing side of our Sun. The Carrington Event, of course, is the most famous happening, and the 1957 episode is one of the least known. All involved a CME that hit or threatened Earth, causing a geomagnetic storm. This event is a major disturbance of Earth's magnetosphere, explains the National Oceanic and Atmospheric Administration (NOAA). It happens when there is an exchange of energy from the solar wind in the space environment surrounding Earth.

What's attention-grabbing is, when using the Planets Today and the AstroGrav systems, I can accurately predict the Mercury/Earth/nearby-star conjunctions. But I can't yet forecast the creation of an active region. Thus, I will, with the use of the STEREO-A satellite data, be able to foresee a possible Earth-striking CME at least three days ahead of time.

It's interesting that the months that don't appear on the Wiki list of solar storms are December and June—the summer and winter solstice months. Why? This may be explained by the fact that the most nearby stars, such as the Milky Way, on or near the ecliptic are the least available for conjunctions in these two months.

I also will consider early September as a high-probability time because two of the several examples happened then. September solar storms happened five times on the Wiki list of solar storms. On a 10-month basis, the average is 2.8 per month. So September is nearly twice as likely to have solar storms affecting Earth as any of the other 11 months. Keep in mind that these solar storm events happen, on average, about once every 15 years. (7)

Segment 5 ("Mercury Event")

Images from theplanetstoday.com with permission from Hayling Graphics Limited

Note the positions of Earth and Mercury are nearly the same angle. Time is in UTC. Mercury and the star HIP 115445, 62.5ly away and 0.1 degree apart were in conjunction on 09/09/1957 at 1924 UTC.]

For now, I'm going to focus on Mercury because it has become obvious that the planet Mercury is at least indirectly connected with solar flares and CMEs. It may lead to an improved understanding of all solar phenomena. My focus has been on Earth, but as a result, it's unclear on what's happening; thus my research continues to be Mercury-centric.

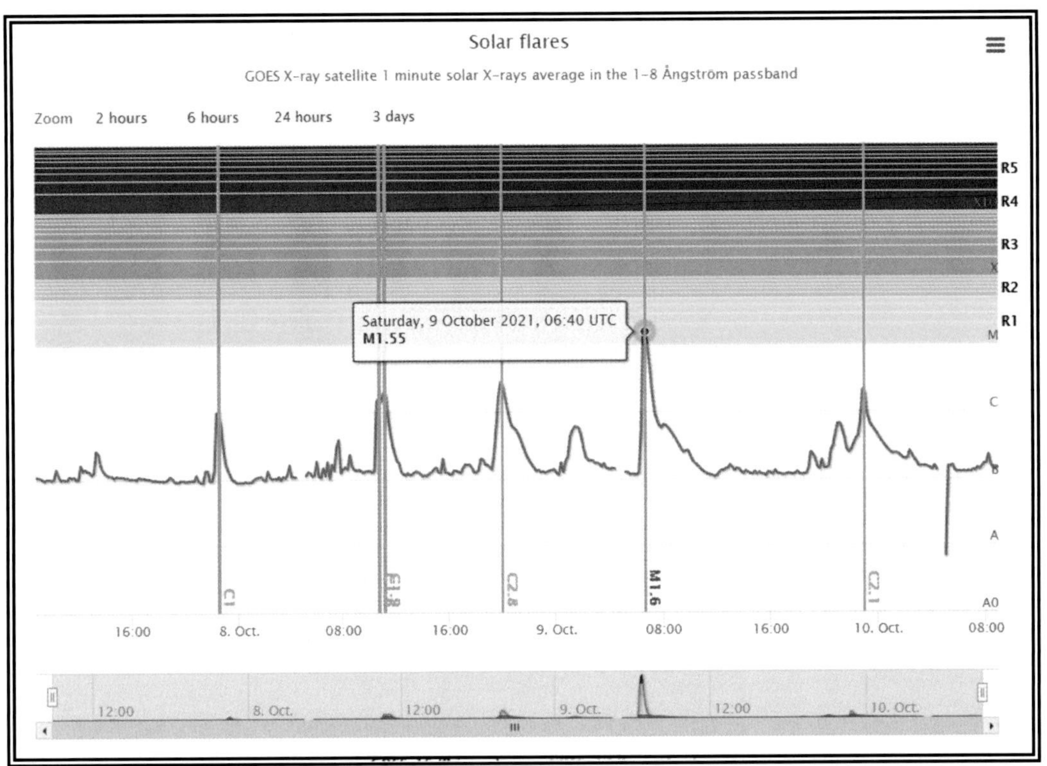

National Oceanic & Atmospheric Administration (NOAA) – spaceweatherlive

On 10/09/21 at 1606 UTC, Earth was in conjunction with Mercury -3.5 degrees apart. The highlighted flare is roughly 10 hours earlier than the conjunction. The flare peaked at 0640 UTC. Now that I see that activity on our Sun will happen some hours earlier than the conjunction and that there may be multiple discharges of static from Mercury, it may be that all five above-C-level flares are due to a Mercury/Earth conjunction. Minutes before the highlighted spike, Earth was in conjunction with WW Psc 918ly away and +0.1 degree from Earth.

This is my first sign of how much the Sun's rotation can shift the planet-planet position phenomena on the Sun from the time of alignment. It will shift the phenomena some hours before the conjunction because the Shaylik is curved ahead by the Sun's magnetic rotation effects on the Shaylik. Earth's rotational shift of phenomena is three to four hours by the diurnal variation of a compass. This is a key factor, possibly for planet-planet configuration I haven't accounted for in the past.

FIVE FLARES EFFECT OF PLANETS AND SHAYLIK

The five flares data became available on Helioviewer (a solar and heliospheric image visualizing tool). This data is evidence of a possible Mercury or Mercury/Sun Shaylik source of X-rays.

Date	Time	Level Location Comment
10/08/21	**1109 UTC C1.9 0,0**	**Unknown Location**
10/08/21	**1950 UTC C2.7 0,0**	**Unknown Location**
10/09/21	**0640 UTC M1.6 0,0**	**Unknown Location**
10/09/21	**2300 UTC C2.0 -38, -32**	**Unknown Location**
10/10/21	**1300 UTC C2.3 0,0**	**Unknown Location**

Segment 6 ("Earth/Mercury Solar Smash")

10/09/2021: "An M1.6 solar flare peaked yesterday at 06:38 UTC. Source of the eruption was sunspot region 2882 which was in an earth-facing position at the time. Type II and IV radio sweeps were observed which suggested straight away that the event should be eruptive and thus launch a coronal mass ejection towards Earth. When coronagraph imagery became available, it became quickly obvious that the solar flare indeed launched a coronal mass ejection towards our planet." (8)

On 10/09/2021 at 1610 UTC, Earth was in conjunction with the Star WW Psc, 918ly away 0.0-degree separation, and Mercury was at -3.5 degrees. Mars was involved 11 days earlier, making an AR in a conjunction with the star HIP 61791, 665ly away and a nearly perfect conjunction at 0.03 degree. This is another example of an Earth/Mercury-related CME with a conjunction star. There are no coincidences—this is another Shaylik happening.

Images from theplanetstoday.com with permission from Hayling Graphics Limited

To cut a long planetary story short, there's a direct correlation of the Mars, Earth, and planet geometry between the first two Earth-striking CME events. This is one of the reasons I believe that planet alignments with large nearby stars is a cause of solar phenomena. And planet lineup events that have happened in the past will likely continue to occur.

There's just one more observation I can make based on this chapter's observations. The location of Earth during these CME events is in the same part of the sky as the Mars global dust storm period described in chapter 3. The months for these five Earth CME events are from July to October. The Mars global dust storms are from Ls 225 to Ls 310, which is equivalent to from Earth's months of July to October (Mars Ls 214 to Ls 290). Mars varies in its elevation from the ecliptic (Earth's orbital plane) by 1.85 degrees, which means both planets are exposed to the same stars on occasion.

So it appears that when Earth or Mars is located between Earth's July-to-October locations, there is a significantly higher probability of a CME coming to Earth or, for Mars, a global dust storm. For Earth, that is if Mars has made an active region about 13 days earlier and Mercury is on the same side of the Sun as Earth and Mercury is having a nearby-star conjunction. I haven't noted any of Mars's precursor conditions for Mars global dust storms other than the star conjunctions noted.

You may think I was thinking ahead when I chose the five instances of Carrington-like events because they fit the time of the Mars star conjunctions and global dust storms. The truth is, I chose the 2017 CME event because it happened while I was doing this research and the locations of Earth and Mars repeated exactly with the Carrington Event. I was, at the time, fascinated by the possibility Mercury could lead me to better understanding of what causes solar flares and picked the other four CMEs based on the position of Mercury in relation to the Earth side of the Sun at the time. I didn't see the coincidence of same locations until I was putting the final touches on this chapter. Also, I didn't use instances when the date was not included because I couldn't then find a possible named star involved with Mercury.

If, in the future, you would have asked, "Did you notice the Earth CME events happened with the same star locations as the Mars global dust storms?" I would have been terribly embarrassed and pleased at the same time. Sometimes I can be as observant as a dead tree stump. My wife, Claire, would surely agree with this statement! This actually created another epiphany moment for me, and once again, I couldn't type on my keyboard.

Meanwhile, moving along (pun intended), let's take an up-close-and-personal look at moving and mind-boggling happenings and Shaylik in chapter 13: "Flow of Shaylik."

NEW THOUGHTS, OLD UNIVERSE CLUES

- ✓ Happenings like the Carrington Event are due to planet/star conjunction irregularities that are within the magnetic field of the Sun, which is powerful and can create magnetic storms that cause damage to electronic equipment on Earth.
- ✓ The only thing in our Universe that could reach out light-years to affect our Sun is magnetics . . .
- ✓ Magnetic forces are said to reach out infinitely from their sources.
- ✓ Mars seems to be an active-region generator, while Mercury seems to be a solar-flare maker.
- ✓ The six Earth CME events cover the same months and therefore stars as the Mars global dust storms.

CHAPTER

Flow of Shaylik

The light which puts out our eyes is darkness to us.
Only that day dawns to which we are awake.
There is more day to dawn. The Sun is but a morning star.
—Henry David Thoreau

Did you know Shaylik (the building blocks of the Universe) flows nonstop like a body of water? Think of the Mississippi River, which is always flowing; even after the strong New Madrid earthquakes in 1811–1812, which caused the river to move backward, it was still flowing. Shaylik is like this analogy because, like the Mississippi River, it never stops flowing.

So no matter what dimension I'm talking about or what natural or man-made form it takes, Shaylik is a form of perpetual movement or flow—or kinesis. Motion of mass has kinetic energy, so Shaylik is always kinetic energy—potential energy when we, or nature, shatter Shaylik like we do with a Van de Graaff machine (an electrostatic generator that uses a moving belt to build up electric charge). Then the Shaylik displays potential energy because it will continue to try to reattach to other nearby Shaylik rivers. It will eventually recombine with something to become kinetic energy again. So a lake or pond of Shaylik isn't possible for any length of time, and it will always find a way to flow like the great Mississippi River, the second-largest river in the North American continent.

Exploring Shaylik and Its Drift

So I could define static electricity as the unorganized, environmentally unstable flow of pieces of Shaylik. Part of the concept of Shaylik is the flow of at least the two-dimensional mass version between stars and planets. I know it's not three-dimensional mass because there would have been proof in the flow. But the description of gravity being two-dimensional mass flowing at the speed of light into Earth strongly suggests there is indeed a flow. But there isn't strong scientific evidence of proof, much like the three-dimensional mass.

I get it that it may be impossible to measure two-dimensional Shaylik. Why? Because no matter what tool is put to work to measure it, our instruments may not be able to detect something so small and weak in force. But if there is a will, is there a way to measure Shaylik? Anything is possible.

Back to my outside-in thinking, I believe the detection of anyons is the closest I know of to measure for two-dimensional Shaylik. (Think of your cell phone's app that displays real-time gravity in G's as well as real-time acceleration.) But that's the result of what Shaylik is doing and may or may not help.

So as far as determining the movement of Shaylik, the best indicator may be what happens to a black hole, which is the ultimate Shaylik gobbler (think relentless Pac-Man slayers). It's known that magnetic lines are going into black holes. It's also known that mass is drawn into a black hole and can pass out from the north and south poles of the black hole. But all this happens beyond the black hole's event horizon, which hides all that is happening from any evaluation we can do. It's back to square one or thinking outside the black hole.

Gauging the Course of Shaylik

The next-best thing to measure the flow of Shaylik is a pulsar known as a celestial object. It's believed a pulsar is a quickly rotating neutron star that emits regular pulses of radio waves and other electromagnetic radiation. Alas, no pulsar is close enough for us to see the detail of what's happening on its surface. This is probably a good thing because the energy coming out of the pulsar would cook Earth if it were close to it and in the right direction from it. The typical pulsar is coined a "neutron star" because it's suspected to be a large ball of neutrons with no electrons or protons to speak of. These structures are also sometimes called a "magnetar" because the very strong magnetic forces it has are huge.

Pulsars are known to draw mass from the nearby space and then spew some of it out its poles. Often this type of star wobbles on its axis, and this causes the energy coming out of the poles to make a cone beam of energy rather than a steady beam. This energy, when detected from far away and in the cone of energy, looks like a ship's lighthouse, flashing at or near the rate of the star's rotation. So we can imagine that not only magnetic lines of force are flowing into its equatorial area, but similar magnetic lines are coming out of the poles with extremely high energy.

- *Here Comes the Nature of Force:* This phenomenon is different from what I have in mind for gravity. These marvels, I sense, are due to the large magnetic lines involved in a black hole or magnetar. I envision three-dimensional mass coming into the equatorial area of the magnetar and then following the north and south–aligned magnetic lines created by the magnetar. This is a major redirection of the flow of the three-dimensional mass that is moving at high speeds toward the magnetar's poles.
- *Faster Than a Speeding Magnetar:* These magnetars are spinning at a high rate—rotating as fast as 1,000 revolutions per second. The three-dimensional mass coming into the magnetar is spiraling

> into it relatively slowly. The difference in speed between the magnetar and the three-dimensional mass coming into it causes what is called the "triboelectric effect." This increases and supports the already-existing magnetic fields that are there. I see this three-dimensional mass, which most likely is in the form of a hydrogen cloud, not only coming into the magnetar but then spiraling up or down on its way to the poles. So it's spiraling up or down at a high speed from all around the magnetar toward the poles. If I were standing at one of the poles, I would see this three-dimensional mass coming at me (like a missile aimed at a target), following near the surface from all directions. All of this matter is on a collision course to where I'm standing. I know this matter can't literally collide with the other matter because it has magnetic fields of its own that won't allow a direct crash because of repulsion by the inverse square law. So they can come close but not physically cause impact. The pressure where I'm standing is huge, and all the mass is again redirected magnetically directly up from me and heading out to space in the form of a beam. (Imagine a laser beam used by astronauts on a spacecraft mission.)

The point? Nothing passes through the magnetar. It's all going around it. If this were two-dimensional Shaylik and this were a planet, it would pass through the planet's mass to its center because there's mostly nothing between electrons and their nucleus. But this isn't a planet; it's a neutron star. There are no electrons surrounding a nucleus, just neutrons packed together. The bottom line: No spaces exist for the two-dimensional Shaylik to go through—what happens with Earth is very different. Atomically, Earth is mostly gaps between atomic structure. Most of the Shaylik coming into Earth from other planets and stars all around it can pass right through it toward its center. A significant portion of the Shaylik will slow due to eddy-current effects and change it from two-dimensional to three-dimensional mass.

Gravity Is a Fragile Force

- *The Force of Gravity:* It's important to note here that gravity is known as the weakest force known to man. It's not that it doesn't have a major effect on things like planets and stars, but its effects are distributed throughout the environment such that, in a specific spot, it is weak compared to the other known local forces. So that little effect on Earth is gravity. But it has a big effect because it's everywhere and its forces are cumulative. All the atoms in and around Earth have their own magnetic fields. Also, all the components of the atom have their own magnetic fields. The two-dimensional Shaylik can pass through and between the parts of the atom without effect, but it cannot avoid the magnetic fields associated with them. The Shaylik is slowed while passing through these electric fields. As two-dimensional Shaylik slows in the presence of atomic magnetic fields, it imparts energy in the direction of its flow on the atoms. This is what is called the eddy-current effect that we are familiar with.
- *A Higher Dimension:* Two things can happen as a result. The Shaylik can change to higher-dimension forms and stay with the other three-dimensional mass, or it can simply pass through the three-dimensional mass and only be slowed in its path. The directional energy of the Shaylik

will either be absorbed completely, as in the case of changing to a higher-dimensional structure or, as a minimum, not change but give up some of its directional energy and continue on.

Is the Universe Restarting?

Until recently, I had no evidence to support the idea that two-dimensional Shaylik coming into Earth could change to two- and three-dimensional mass. Now I have some indication this can and does happen—and the future may surprise us all.

So with all this outside-in thinking, how can I explain the planets orbiting our Sun? There are essentially only two forces involved. The one we are all familiar with, called "centrifugal force," is the obvious choice for the force holding planets away from the Sun. This is a well-studied fact like the theory of evolution. The other is gravity, but not for the reasons we currently think. We currently say Earth is attracted to the Sun by gravity. I say, because there is no attraction magnetically, the planets are pushed toward the Sun.

- *Here Comes the Sun:* Think about the Sun as if it were a rare silver dollar—not surrounded by planets. It's just there, surrounded by the stars of our galaxy. In this thinking, those other stars may have planets or not. Now imagine all those millions of stars all around are having two-dimensional Shaylik flowing from the other stars to our Sun. Since Shaylik by nature is two-dimensional, those lines of Shaylik will be very straight. The size and effect of those lines will all be different depending on whether the particular star is close or not, or large or small compared to our Sun. One other thought is that the flow direction is based on the size of the star. That is, if it's more massive than our Sun, then the flow for that star will be away from our Sun. When the star in question is smaller than our Sun, the flow will be toward our Sun. Right now, by far the majority of the nearby planets are smaller than our Sun, so the dominant flow is toward our Sun.
- *Walking on Sunshine:* Now put an orbiting planet in the picture. This no-name planet is then smaller than our Sun, and the Shaylik from it is going to our Sun. In the planet's path around our Sun, it passes through the Shaylik from the other stars, which will cause a minor and temporary dragging effect. Most of the time, the planet is exposed to Sun-incoming Shaylik from outside our Solar System and, through the eddy-current effect, is pushed toward our Sun, countering the centrifugal forces such that the forces are balanced and the planet just keeps orbiting our Sun.

Dimensions and Shaylik Simplified (Sort of)

This theory, if valid, means that in the big picture of what's happening in our Universe, the big just gets bigger, more massive, and the small just gets smaller, more energetic. But note that all the high-energy phenomena we're aware of is small, such as gamma rays, X-rays, and photons. So what would the mechanism be that supports this idea? Read on.

- *Two-Dimensional Shaylik:* The predominant two-dimensional Shaylik configuration between masses is the double-helix. I believe that two ropes twisted around each other are unable to get close to each other because of the magnetic repulsion between them. One other feature is, the two ropes are different in diameter. One is fatter than the other, and this is significant in that it indicates they are different is some physical way—still two-dimensional Shaylik but not the same.
- *Flow of Shaylik:* I see the larger-diameter rope as the flow of Shaylik to the larger mass, which eventually becomes three-dimensional mass when it enters the concentrated magnetic environment of the larger planet or star. The smaller of the two ropes is more like energy than mass.
- *Three-Dimensional World:* Still, I evolved to explore this rare environment. But my world is an illusion in terms of energy. Energy is the basis for all that there is. The atoms that make up my body have more stored energy by far than the biggest bomb we have ever made. This is true for all matter that is three-dimensional. So my atoms have a lot of stored energy, and that energy is in the form of folded-up two-dimensional Shaylik—this is why Albert Einstein said that energy is mass and mass is energy. The realm outside my own is a realm of energy, not three-dimensional matter. I also know that energy can take many forms. Just consider the environment I am in right now. There is a huge variety of structures here, and this world is only a small sample. So there probably will be a variety of structures that are not three-dimensional in the two- and possible one-dimensional world out there. Hopefully fewer varieties than are here on Earth.

So I see Shaylik in the double-helix structure as versions of energy. One version is more like pure energy in that its energy is easily absorbed by three-dimensional masses. And that means it mainly heats three-dimensional masses. This could be as simple as what we call "static electricity." After all, in an environment that has stellar winds from a star, each nearby planet will develop a static charge based on its size and proximity to the star it surrounds.

Earth must have more static charge than our Moon because it's larger and has an atmosphere that could act as a capacitor. A capacitor is a device in our electronics that can temporarily hold a static charge. The Moon has little in the form of an atmosphere, so it's likely to be less statically charged than Earth. The result: It would be that the greater static charge of Earth will try to find a way to discharge to a location with less charge. If Shaylik is a superconductor, then it will be the path of least resistance to the Moon or other smaller planet nearby. I believe this is what's happening in the smaller rope of the double-helix structure between Earth and the Moon.

What's more, this is Shaylik that is less organized and more fractured than the Shaylik in the larger rope of the double helix. It's made of smaller pieces all moving toward the smaller structures. This Shaylik, for the most part, cannot become three-dimensional mass but can join with it. The larger rope of the double-helix is more like a laser beam. Its structure is stranded, is coherent and all working together as a system of parts bound with each other. This is the structure that can become three-dimensional matter if it is going to the right environment. So the larger rope of the Shaylik double-helix is physically more uniform but made of the same basic two-dimensional matter.

This thinking would apply throughout the cosmos, all with the same flow characteristics. If two masses are equal in all ways, there is a Shaylik double helix. between them but it's flow will not be the same intensity. The odds of this kind of equality of mass are slight. There are too many environmental variables involved to allow this equality.

- *Moon Matters:* Our Moon's surface is tidally locked in its orbit around Earth. This means one side of the Moon is always facing Earth. One more fact: The thickness of the Moon's crust is thinnest on the side facing Earth. There is one other reasonable explanation for this I have seen. To add to the list, I think because the Moon is smaller than Earth and the Shaylik between Earth and the Moon converts matter from the Moon to two-dimensional Shaylik and passes it on to convert back to three-dimensional mass in the earth. This two-dimensional mass comes from the three-dimensional mass of the side facing Earth. This thins the Earth-facing crust. Three-dimensional mass is converted to Shaylik and moves energy from Earth to the Moon, which heats the Moon and causes occasional bright spots on the Moon that our amateur astronomers have noted for years. This also explains why the double-helix that is the configuration of Shaylik between masses is not uniform in size. One of the two parts of the double-helix is always smaller than the other. So the side of the Moon always facing us has been losing mass while the opposite side has not, and that is why the outer layer facing Earth is thinner.

  At this point in my research, I think the smaller rope of the double-helix Shaylik is unorganized two-dimensional Shaylik and the other larger rope is organized two-dimensional Shaylik. The larger Shaylik is in an assembly that is continuous strands of two-dimensional Shaylik that has common chirality and common frequency. This coordination is why it has a bigger magnetic field. All the magnetic lines of force are supportive, creating a stronger field .

  On the other hand, the smaller of the two ropes is akin to and maybe what we know of as static electricity. This is Shaylik that has been magnetically torn to pieces. This happens whenever a three-dimensional mass like Earth or our Moon moves quickly through the solar winds. Some of these masses have an atmosphere that will support or increase this frictional tearing of magnetic lines. The object's size will be the key contributor of static-electricity generation. Earth will develop a larger charge than the Moon just because it is bigger and has more surface area involved. Also the distance between the Moon's center of mass and Earth's has been increasing over the eons, which could be because the Moon is getting lighter and Earth getting heavier over those same eons.
- *Static Electricity to Gases:* The larger-charged mass will have a larger amount of static electricity. When there are differing amounts of static charge, the larger charge will try to balance the charges between the two masses. Static charge will flow to the lesser-charged mass. So there will be a constantly varying flow of static electricity from Earth to our Moon. It varies because the solar

winds vary in density and speed. This will be true for all masses that are in an environment that is not moving at the same speed and direction as the mass itself. This is the more common condition because everything in the Universe is moving relative to the other nearby masses or gases. Seldom are the gases moving exactly with the masses near them . . . We cannot detect the flow of static electricity directly because the Shaylik pieces are just too small. What we can detect is what the Shaylik can become, which are electrons. So our science sees this as a flow of electrons when it is actually more complicated than that.

Shedding Light on Shaylik

So the two ropes of Shaylik are different sizes because they are organized differently and have different magnetic intensities. A laser beam of light doesn't spread out much because the light from a laser is what they call "coherent light" that's all working together. Now imagine a laser beam that is coherent, wrapped around a beam that is not so organized, keeping the disorganized one together. It's acting as a shepherd (or an Australian shepherd on the job), keeping the herd together. It's just a matter of perspective. Earth orbits the Sun or the Sun shepherds Earth. They both have the same result; it's just a matter of perspective.

- *Jumbled Shaylik:* The thing is, disorganized Shaylik is passing to the smaller mass and maybe becoming electrons. Organized coherent Shaylik is passing to the larger mass and becoming electrons, protons, and neutrons—if this is the case, then when we establish a Moon base, we may have an easily accessible source of electricity. More than likely, there would be concentration points on the Moon where the static electricity is greater than the Moon's average static level. These points would probably have to do with any iron meteors buried in the Moon's crust. We could put static collectors (which are called capacitors) in these locations and bring the energy to another location with wires where the Moon's surface potential is different from the focused locations. This would yield a continuous source of direct-current electrical power for our facilities on the Moon.
- *Shaylik Chaos:* The iron meteors would be around the Moon's limb and beyond as seen from Earth. This is because Earth would get those impacts before the Earth-facing side of the Moon would. Earth acts as a shield blocking those that come to it in the direction of Earth. Static electricity is the unorganized, environmentally unstable flow of pieces of Shaylik. The two magnetic ropes of the double-helix are different for a reason. One is more like mass; the other is more pure energy. And this results in change.

In a nutshell, the long-term future of the Universe, thanks to the topsy-turvy disorder of Shaylik, is fewer and larger stars and more small energetic particles. So as the Universe is constantly changing, you can expect more unusual events to happen around the globe and Outer space.

Meanwhile, unpredictable Shaylik continues to wow us as the Mississippi River did and may do again one day. In the next chapter, find out about some more wonders of the fundamental unit making up all that is in the Universe. You may be surprised.

NEW THOUGHTS, OLD UNIVERSE CLUES

- ✓ The only thing I know of in the Universe that could reach out light-years to affect our Sun is magnetics . . .
- ✓ Magnetic forces are known to reach out infinitely from their source.
- ✓ Mars seems to be an active-region generator, while Mercury seems to be a solar-flare generator through the flow of Shaylik.

CHAPTER 1

Fractal Explained

Learn how to see. Realize that everything connects to everything else.
—Leonardo da Vinci

So what is a "fractal" anyhow? The simple definition is, a fractal is any pattern that, when seen as an image, produces a picture, which when zoomed into will still make the same picture. It can be cut into parts that look like smaller versions of the picture that was started with. According to history, the word "fractal" was coined by Benoit Mandelbrot back in 1975. It's a Latin word that means "broken" or "fractured." An easy-to-understand metaphor is a tree that branches into smaller branches that all look the same except for size. Fractals have many practical uses that are interesting. (1)

It's time for a mathematical puzzle (sort of). Go ahead—look at fractal structure in its plainest form. It may be the square. If you begin with one square and then start adding squares of the same size next to the first square, when you add the eighth square, you have another bigger square. And there's more . . .

Now if you add 16 same-sized squares to the nine-square assembly, you get another even-larger square. This sequence can go on and on and still will create another greater square. The result: A reproduction of a small structure into a bigger structure that looks the same as the small structure is something Mother Nature is good at doing.

The Number Mystery

So it's known that some shapes can be reproduced by just adding more to the first one. Plus, I see this is the case for Shaylik. Its most stable form, the double-helix, can be any size it wants to be. Note the growth to recreation of the original form is in a sequence of steps to completion, so it's done in a quantized fashion. In the case of the square, the quantities to completion by difference are 1, 8, 16. We can also sequence the sums, which are 1, 9, 27. It hasn't escaped science that there were, in this case, three steps to completion and 3 times 9 is 27. Clearly this is a quantized sequence.

A number that keeps showing up in scientific formulas is a mind-boggling attention-grabber. It's strange that the formulas won't work or make sense without this number, and scientists don't know why. Also, it's a dimensionless number, which means they have no idea what it represents. Yet it's constantly used to make the formulas predict reality, which does actually make sense. It's magical. It's like mathematicians know that if you think your formula is correct but it still doesn't give the expected result, then throw this number in because sometimes it fixes the problem.

This magical number is called the "fine-structure constant"—scientists define this term, as in physics, as a fundamental physical constant that quantifies the strength of the electromagnetic interaction between elementary charged particles. Back to the number that equals about 1/137 of something. This is then an adjustment to make something larger when it's used as a multiplier or smaller when it's used as a divider. (2)

This number mystery is just weird and fascinating at the same time. Our normally logical scientists use magic at times! And why do they call it the fine-structure constant? What leads them to this name? I know what "fine" means. I know what "structure" means. I know what a constant is. So I guess it's a small piece of something that is apparently an important part of our world. It's important because it makes things work, or at least makes our math work.

Adding Up the Theory of Relativity

This number is used in math that describes the theory of relativity with its electromagnets, and magnetic flux quanta. A hydrogen atom's electron orbits at $1/137^{th}$ of c (the speed of light). It's also in quantum mechanics: the minimum quantum differences in light absorption and emission and vibrational coherence of electron and photon wavelengths.

Also, it's related to the golden ratio, and fractals and the Fibonacci numbers. Wow, there are a lot of areas in science where it's needed. It seems as if it's related to almost everything from light to atoms to the Solar System. It also seems that it's fundamental to the Universe. That is, it's a part of everything we know of and probably everything we are unaware of. A single thing that the whole enchilada is made of comes to mind.

If it's a structure, as the name implies, then it is something we can describe; it's actually a physical thing that is the basis for all other physical things. It's the ultimate Lego—the toy metaphor of building blocks of the Universe.

So the question is: *Could this be the smallest form of two-dimensional Shaylik?*

I've already shown that Shaylik could be affected by X-rays through absorption or reflection. This is a hint as to the size it could be. X-rays are short in wavelength. X-rays have a wavelength in the range of 0.01–10 nm (nanometers). One nanometer is equal to 0.00000003937 inches. This seems really small when

I say it's less than a millionth of an inch. If this is the case and it can interact with Shaylik, Shaylik must be around this size or it wouldn't be reflected or absorbed.

As a result, Shaylik and the fine-structure constant could be manifestations of one and the same thing. And *if* this is the case, then Shaylik has a dimension of length and is actually a physical thing. Apparently, there is some quality of Shaylik that makes it undetectable by scientists, unless anyons are Shaylik also. In that instance, we have detected Shaylik and are in the dark and don't understand its significance.

Fractal and Shaylik—Mind-Blowing

Ah, the puzzle of the fractals (plural of the word "fractal") of Shaylik. Everything is strained or filtered by the local environment simultaneously everywhere—it's a matter popping in and out of existence in space. So now I see Shaylik as a substance we were previously unaware of that is the basis for all that there is, was, or will be. In its basic form, it's a one-dimensional piece of matter that is smaller than a millionth of an inch. It has characteristics that allow it to combine with other Shaylik to become something larger, like magnetic flux rope. That means this originally one-dimensional piece of magnetic mass can assemble into a magnetic rope, which technically is a three-dimensional structure but still maintains the ability to pass through other stable types of three-dimensional matter. Another and more stable form it can take is a double-flux-rope configuration that's twisting and is twisted around one other flux rope. But amazingly, all is still able to pass through other matter with relative ease. This is called a "double-helix configuration." I see all three of these structures as twisted and flowing between masses like planets and stars. Also, what is mind-blowing is, it flows at the speed of light.

Some scientists say that the vacuum of space is actually full of matter being created and usually immediately disappearing through self-annihilation. The words they use are "popping in and out of existence." This is happening constantly, according to them, and it's observable in a vacuum. This very well could be one-dimensional Shaylik trying to assemble into a higher-level form, which will or will not continue to exist depending on the environment it's in. The three forms or structures of Shaylik I discussed in the previous paragraph are the ones that are stable in our current cosmic environment.

Speaking of the environment, this fine-structure constant happens during what controls the favored configurations of Shaylik. On our planet, a favored configuration is the double-helix, which is a fractal growth of the basic structure of spaceborne Shaylik. Other environments (perhaps on other planets) may have the same and/or other favored configurations. The possibilities are as varied as the environments.

Interestingly, in the presence of normal three-dimensional matter, like our planet, the Shaylik can exist in any of its three basic forms or a higher-order form such as electrons, protons, and neutrons and their assembly as normal matter in the form of molecules and substances. The two-dimensional Shaylik ropes will curve onto themselves, creating the electrons, protons, and neutrons and their related substances that we are familiar with. So the electrons, protons, and neutrons are developed from two-dimensional

Shaylik that had curved into itself like the nucleus of a biological cell but smaller. But why does the two-dimensional Shaylik curve onto itself?

The simple answer is, because of the magnetic environment of a mass like Earth. It's highly concentrated and, through magnetic pressure, forces the two-dimensional Shaylik to bundle up like balls of thread or nuclear DNA. Anything being produced this way that isn't stable in the environment will immediately pop out of existence—poof!—and be gone just like that! The stable ones, however, will remain as new matter.

OK, I see the basic structure that's the most stable and pervasive is the double-helix, which is stable because it's in a circuit between masses. This is a fractal spinoff of what I see as the basic structure of Shaylik. The basic structure is a single piece of a one-dimensional magnetic line that is free to roam (think of a wild coyote in a desert) the cosmos. I imagine it as a piece of string that is twisted and is two-dimensional. We see this version of Shaylik in many ways and call it a photon, or a gamma ray, or a neutrino, or light or infrared radiation, or static electricity.

My son Jack and I have talked a lot about the fractal concept as it applies to the cosmos many times. We now see that structures like Earth are fractal derivatives of the basic Shaylik structure that has morphed billions of times in billions of directions, which were controlled by the environments they developed in. So yes, there can be and probably are more Earth-like planets out there. They would have subtle differences but, for the most part, be familiar to us. There are also planets out there that we could not imagine in our wildest dreams. The possibilities are not limitless but are limited to what Shaylik can do, which is unimaginable.

One of my favorite thoughts is finding a planet that is very much like Earth but has a beautiful feature. The feature is ball lightning that can last days instead of minutes and is present in all the colors of the rainbow. This planet would have no night because these multicolored balls of light would make a beautiful light show at night and float all around us. If we touched them, they would disappear, but it is taboo to touch them and is just not done except by accident.

Dimensions and Shaylik Dissected

Shaylik is the one-dimensional basic structure that all we know of, all that was and all that will be, is what it's based on—no ifs, ands, or buts about it. Here are some related facts to digest:

- It's possible that through quantized fractionalization, two-dimensional Shaylik can become a form that is repeated in quantized steps from the smallest to the largest of structures in the Universe. It also may be that two-dimensional Shaylik is the basis for quantum physics.
- The fine-structure constant may be a key to understanding the phenomena of Shaylik.

- There may be no limit to the forms and thereby functions that Shaylik can take and be stable if the environment allows.
- And during the roaming around the cosmos, it can make and be its own environment much like the gas hydrogen. It can be a gas or a liquid, a massive cloud in space, and even a star. I also imagine it as a piece of string that is twisted and is two-dimensional. We see these versions of Shaylik in many ways as what we call a photon, or a gamma ray, or a neutrino, or light or infrared radiation, or static electricity.

So it may be that the fine-structure constant is actually just a mere ratio of the number of basic Shaylik pieces in the smallest double-helix versus the number of Shaylik pieces in the second-smallest double-helix. This could possibly be the ideal model for finding quantized structures.

As this theory about fractals requires further research, there is another hypothesis that has been studied, and you have likely heard of it—a lot. Let's go back, way back to the belief of Darwinism and how it has evolved.

NEW THOUGHTS, OLD UNIVERSE CLUES

- ✓ Think of fractal as a pattern that, when seen as an image, creates a picture like a tree that branches into smaller branches. Remember this simile to help you understand its purpose in the Universe.
- ✓ Shaylik (think of a talented magician) can be any size it wants to be. Don't forget—Shaylik and the fine-structure constant could be one and the same.
- ✓ The basic structure is the double-helix, which is stable because of its circuit of masses. This is a fractal spinoff—basic structure of macro Shaylik.
- ✓ Shaylik is the one-dimensional structure, but keep in mind, it also may be two-dimensional. Magic again!
- ✓ And yes, Shaylik could be the ultimate model for finding quantized structures.

PART 6

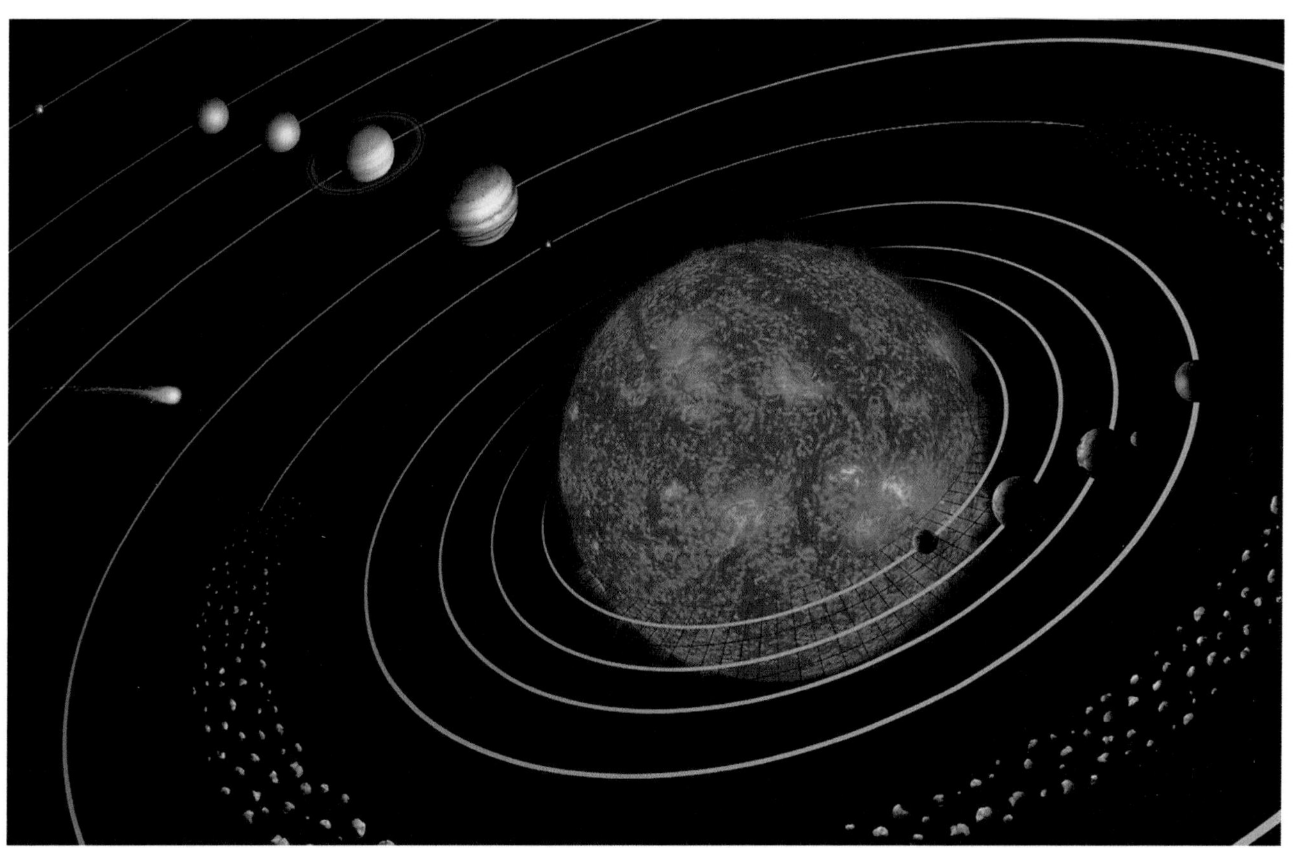

AS THE WORLD'S THEORIES TURNS

CHAPTER 15

Darwin's Theory of Evolution

False facts are highly injurious to the progress of science, for they often endure long; but false views, if supported by some evidence, do little harm, for every one takes a salutary pleasure in proving their falseness.
—Charles Darwin

Welcome to Darwin's Theory of Evolution 101—the stuff kids are taught in school. It goes like this. Charles Darwin's theory of evolution is based on natural selection of God's creatures to survive as the weak ones die, like not being able to make it stranded on one of those survival-island TV shows with challenges. He explains that nature has a way of cherry-picking what has the desirable characteristics to survive. In Darwin's nutshell: Life goes on for the "survival for the fittest." (1)

No theory I know of has been more supported by fact than Darwin's theory of evolution. As far as I am concerned, it should be called "Darwin's Truths of Evolution." Everything he said, including pointing out that man originated in Africa, has been verified. The results of his insight have been rippling through our science ever since. Darwin did not know about DNA, but he did think that someday we would able to describe the family tree of each species. Amazing stuff!!

Darwin didn't know how far back in time his theory would apply. He knew that nature kept trying different things over long periods time—different things at different times and in different environments. He knew that the changes that did not work for one place and time could work for another different place or time. He also didn't know how or what life all began—there were unanswered questions. All Darwin studied were nature's successes because all the failures were gone; they could not survive where and when they were tried.

A New Look at Darwin

We now know that the first living cells were most likely primitive prokaryotic-like cells and Darwin's theory was biology based. What if this is not really the bottom of evolution of matter but just one step in a larger sequence? After all, biology is based on chemistry, which also could be based on something else. And that something else could be the basic form of Shaylik.

As a result, things may have gone like this. Two-dimensional Shaylik that was just a single-helix form was what the primitive cosmos was made of. This Shaylik was also only one chirality and one frequency. This circumstance would allow conglomeration of the primordial Shaylik. This conglomeration would allow the existence of clumping and higher-magnetic-pressure areas. This is a new and uneven environment, which allows for different things to exist than existed before.

This fresh environment may allow electrons to be stable. After enough electrons are created, the magnetic pressure is even higher and protons and neutrons are created. And now the electrons, protons, and neutrons can combine into a hydrogen atom held together by a new kind of Shaylik, the double-helix form. This form of Shaylik can surround the atom and protect it from collision with other particles by magnetic repulsion. It can also keep the electron in orbit around the nucleus.

Once hydrogen is stable, it can become a star and start creating helium, which we are aware is happening in our own Sun. From here on, we have evidence that supports our conclusions as to how all the elements and subsequently life were developed. Is this what happened? There is no way to know as of now.

It's possible evolution of the cosmos began with two structures that were mirror images of each other (imagine two primal magnetic Legos). It would be interesting to talk with Charles Darwin face-to-face in the twenty-first century and listen to his up-to-date spin on his original theory of evolution.

The Evolutionary Universe Surprises

Speaking of Shaylik trying things to see if they are stable, my son Jack and I have thought about the fact that scientists think there's no such thing as a void or vacuum anywhere. They say that, as a minimum, there are virtual particles being created and then self-destructing all the time in what was thought to be empty space. This is an observed phenomenon, according to scientific research.

We believe that this is probably true . . . *and* also believe Shaylik is trying different configurations to see if they are stable and can be built upon to make more-complex structures. Like father, like son, we also agree that even without the popping in and out of virtual particles in a vacuum, there are magnetic lines of force filling all that we call voids or vacuums. That said, there is no place in our cosmos that doesn't have something in it even if it's simply a weak magnetic line of force.

The ball lightning (an unexplained phenomenon) is what we think is an example of a Shaylik structure that is nearly stable but cannot survive in our current environment for more than an insignificant amount of time.

To wrap up this chapter, the Universe is like the four seasons on Earth and climate change. Likely it will continue to surprise scientists, and you too, as different events happen on Earth and in space. Not only are changes amazing, but when discoveries are made, it's the knowledge of power that shows us the power of the mankind-cosmos connection.

Now that you know some of my thoughts about Charles Darwin, what a perfect time it is to bring up the next well-known scientist from yesteryear. I can't leave out math guru Albert Einstein, but I can add a few notions of my own to widen the circle of his theories.

NEW THOUGHTS, OLD UNIVERSE CLUES

- ✓ Charles Darwin, an evolution scientist, believed life is for the survival of the fittest.
- ✓ Darwin also thought that we would be able to describe the family tree of species.
- ✓ It's possible evolution of the cosmos began with two structures that were images of each other, like two magnetic Legos.
- ✓ Scientists have faith in that there isn't such a thing as a void anywhere. As a minimum, there are virtual particles, they suppose, being created and then self-destructing all the time into empty space.
- ✓ The ball lightning (unexplained phenomenon) is an example of Shaylik—structure that can't survive in our environment for a lengthy time.
- ✓ Like it or not, the Universe will continue to surprise scientists because of the evolutionary cosmos.

CHAPTER 16

Rethinking Einstein's Gravity Theory

There is grandeur in this view of life, with its several powers, having been originally breathed into a few forms or into one; and that, whilst this planet has gone cycling on according to the fixed law of gravity, from so simple a beginning endless forms most beautiful and most wonderful have been, and are being, evolved.

—Charles Darwin

Albert Einstein, more than any other individual, made us aware of the power of mathematics. This was his special ability. His example forced our science toward math-based thinking. We came to believe that if the math says something will happen or could be, it must be true. I'm sure he knew that math is a tool and that a tool used correctly will help accomplish the task at hand. But like any tool, it can be used incorrectly and lead us in the wrong direction.

Einstein's gravity theory made a splash in the past and still is debated in present-day in science classes. In general relativity, there's the phenomenon of gravitational time dilation: Roughly speaking, clocks in the vicinity of a mass or other source of gravity run more slowly than clocks that are farther away. This phenomenon is closely related to the gravitational redshift. Read on—and I will explain the theory to you.

Gravity and the Environment

As I move from above Earth toward its center or down, I'm moving toward a denser magnetic environment. Not only the three-dimensional mass is getting more and more dense, but so is the incoming Shaylik. The Shaylik is being forced into a magnetically more-restrictive environment, causing its vortex shape and compressing its lines of force together. In fact, all the magnetic forces are being compressed into a smaller area, making the all-magnetic forces more dense.

We use our most accurate clocks to show the effects of time dilation. These clocks are based on atoms like cesium. In order for these clocks to perform their best, we must keep them at the same temperature. This is because temperature or energy level will affect an atomic clock's accuracy. So scientists try their best to keep environmental differences from being the reason for differences between two clocks of the same design. They can and do control temperature of these devices as best as humanly possible.

What's more, audio vibrations can also cause heating or cooling of the atoms, so they should be controlled also. I'm sure they would like to control the magnetic environment also, but that is a whole different and tougher problem than temperature. Magnetic lines of force can pass through any structure we can devise. I'm not talking about electromagnetic forces like in the microwave you use. We know how to stop them with simple conductive structures. I'm talking about the magnetic forces like neutrinos and—of course—Shaylik. These aren't controllable and will change with the changing magnetic environment.

Shaylik at Relative Work

The neutrinos are different from Shaylik in one important way. They are traveling through space independently of any other matter or other neutrinos. They really do and can just pass through Earth without any effect on Earth or the neutrinos themselves. It's as if Earth is not there, nowhere to be seen. Gone.

Shaylik is a collective-force system. All Shaylik can work with is a single mass. If all the Shaylik from all the sources of Shaylik were the same frequency and chirality, then there would still be a compression of the Shaylik as it comes into Earth because of the different magnetics of Earth's materials repulsing the incoming Shaylik.

But it's unlikely that the Shaylik coming into Earth, heading for its center of mass, is the same in frequency and chirality. Why is this nearly impossible? All of the stars and all of the planets would have to be exactly the same. This isn't going to happen.

So the density of Shaylik will increase as it gathers toward Earth's center of mass. This change in density of one-, two-, and three-dimensional mass due to higher and higher magnetic pressure will have an effect on atoms. It will slow the orbital period of the electrons due to increased magnetic friction from the magnetically denser environment. These are eddy-current effects that are the basis for the theory of general relativity.

Research suggests in special relativity: From the perspective of a viewer, a moving clock moves slower than a same-built clock at rest. The rest of the atomic processes moving nearby the clock are slowed down in a matching way.

Moreover, when two observers "speed past each other, each will find that the other's clocks go slower." Some features of this result are explored in the "dialectic of relativity"—a geometric analogy is presented in time dilation on the road. (1)

Under special relativity, the same thing will happen. The density of the magnetic lines, from the one-, two-, or three-dimensional magnetic lines the clock and rocket are passing through are more dense per unit time. It's like the magnetic fields external to the rocket and clock on board are compressed, which slows the atom's electrons due to magnetic friction and time slows.

When the rocket is still on the launch pad, there is no apparent compression of the magnetic fields around it. So the atomic clock remains the same as those around it. So if Shaylik increases briefly, time will slow temporarily, and if Shaylik decreases briefly, time will speed up temporarily. The interesting thing is, we would never know it because all our clocks are doing the same thing when they are in the same environment.

So I—and you too—am made of the same atomic structure as the clock. Electrons can slow and speed up for the same reasons. Thus, hypothetically, I would age more slowly than those people I left on the launching pad. But if I never return, I'll never see the difference!

Einstein's Theory Hits Home (Sort Of)

This is absolutely the simplest explanation of Albert Einstein's general and special relativity that I have ever heard. Also, this means that general and special relativity time dilation can't explain gravity because it's a result of gravity, which is not the cause. And time dilation is a result of an environment's magnetic density and indirectly related to gravity though coincidental effects, and it is a mechanical phenomenon.

So Albert Einstein's understanding of the effects of gravity was spot-on. As far as I know, he never explained how gravity happens aside from saying it happens where there is mass/time dilation.

OK. You've got my news flash on how gravity works in the Universe, but don't forget theories will change as the clock moves forward in time. You'll discover in the last chapter that theories about Earth and Outer space will evolve as time passes in the cosmos, decade by decade, and century by century.

NEW THOUGHTS, OLD UNIVERSE CLUES

- ✓ Einstein made us aware of mathematics and its power to explain the effects of gravity and the Universe.
- ✓ Sure, Einstein's gravity theory isn't perfect, but it made an imprint and is forever in science books, which can and do include findings that grow and develop as we go forward.
- ✓ Shaylik can't be controlled because of the changing magnetic environment on Earth and in the Universe . . .
- ✓ And remember, Shaylik is a collective-force system.

- ✓ It's unlikely that the Shaylik coming into Earth, heading for the central mass, is the same in frequency and chirality. Why? The stars and planets would have to be the same. It's impossible due to change.
- ✓ The general and special relativity time dilation can't be explained by gravity because it's a result of gravity (Shaylik flow). Time dilation is caused by environmental magnetic density and the subsequent eddy-current effects on matter and simply indirectly related to gravity.

CHAPTER 17

The Evolution of Our Universe

The world will not be inherited by the strongest,
it will be inherited by those most able to change.
—Charles Darwin

Since I was a kid, I've witnessed scientific theories be debunked, confirmed, and changed up to open my eyes and make me wonder, "Whom do I believe?" And often, as you can see in this book, I create my own theories about the cosmos. As the Universe evolves, there will be more Darwins and Einsteins working on their hypotheses to prove their vision is the right one to believe about the evolution of the Universe.

Shaylik and the Cosmos Connection

Not again? I'm amazed. One of the most exciting things that keep happening with this theory is when I find new confirmation of one or more of the concepts herein, like when I had been thinking that two-dimensional Shaylik can bundle up when it is entering a mass like a planet to become two- and three-dimensional matter. In the beginning, I had no evidence to support this thought, but it would explain why Shaylik coming into Earth is not also coming out. It had changed and made Earth a little more massive. So it never leaves; it changes form. Then, in 2020, I found out about anyons, which support the existence of one-dimensional and two-dimensional matter. I was so excited I couldn't type on my keyboard.

Well, it just happened again. In early September 2021, I read in my copy of *Science News* magazine that there is an existing theory, the Breit-Wheeler process. It hints at the creation of electrons and positrons through the collision of light particles. By the way, this theory was developed 10 years before I was born. Since I see light as just a piece of Shaylik, this was again a time I couldn't type on my keyboard from the excitement.

The article supports the idea the Shaylik could turn into three-dimensional matter under the right environmental conditions. It comes from the scientific world of quantum electrodynamics. (1) (2)

So our scientists made a machine that can accelerate a gold atom's nucleus to very near the speed of light. A gold nucleus has 79 protons and 118 neutrons all in a bundle that manages to stay together during the test. As it speeds up by magnetic forces, a cloud of photons appears around the nucleus and it gains its own strong magnetic fields that are perpendicular to the machine's magnetic forces. So just so far, we know from this that magnetic fields can create photons under these environmental conditions.

Once the nucleus and its surrounding cloud of photons has reached the speed goal, they let it loose, heading for a target gold nucleus that has the same photon cloud and magnetic fields. The goal was to have the two photon clouds collide but not the nuclei. They apparently accomplished their goal: Photons collided, and electrons and positrons were produced. This was in accordance with the expectations of the Breit-Wheeler process. So now we have light colliding and producing electrons and positrons. Now we are getting three-dimensional matter from what is seen as two-dimensional matter. This was certainly good news to me.

In the article, they wonder if this is real matter because they see it as "virtual particles." Particles that aren't stable and quickly disappear are what they call virtual particles. The neutron disappears in 15 minutes when it is not in an atom's nucleus and they do not call it virtual. So I guess it's a virtual particle if it quickly disappears instead of slowly? There are probably many environments where the production of electrons and positrons with colliding light is stable; we just have not reproduced the required environment. Maybe Earth has done it?

Theories from Unexpected Places

I knew from the beginning that if I came up with the right concept, I would keep finding delightful support from unexpected places. Also, I expected a lot of coincidence situations, situations where everything seemed to fit, but it just did not happen often enough to claim as supportive evidence. This has been happening and will continue. And it will keep me going until I can't anymore. My daughter thinks I am weird because this is my fun. As I recall, she mumbled something about geeks!

They don't mention any theory as to why the photons appear in this situation. I guess this is common knowledge I didn't possess. I do have an idea on how this can happen, and it relates to static-electricity production.

The condition inside their machine is of course a strong magnetic environment. So I see that the magnetic fields of the gold nuclei are interacting with the magnetic field of the machine. I think of this circumstance as magnetic friction, just the same as rubbing my hands together. The atoms in my hand are not rubbing together, but their magnetic fields are.

This is similar to the way a Van de Graaff machine works. It works by a rubber-like belt passing by, but not touching, a conductive metal structure. The metal structure is connected to a ball or other structure that is statically charged. If the reader has not experienced a Van de Graaff machine, they should try one—it's fun and educational.

The fact that the conductive metal structure is not touching the belt is a key point. The only connection between the moving belt and the conductive metal is magnetic. The magnetics of the belt and the metal are interacting with each other to generate static charges. This works because the belt is a nonconductor of electricity and the metal is a conductor and both have magnetic fields around them.

These magnetic fields are being shredded by the continuous collision of the magnetic fields of the belt and the metal. This creates pieces of magnetic fields that are detached from the atoms they surrounded. These detached magnetic fields are loose and free to roam the area around the Van de Graaff machine. If the machine is turned off, they will be absorbed by the air and structure around the machine. But this takes time to happen, so the effects seem to linger for a time. These are pieces of Shaylik I'm talking about. I guess you could call them virtual particles because the disappear quickly by absorption into the surrounding environment.

Under certain environmental conditions, this static electricity (pieces of Shaylik) can last for a while and even cause your socks from the dryer to stick together. This is an example of two nonconductive materials experiencing static due to their atoms' magnetic fields being forced by each other and the air in the dryer. Under the right environmental conditions, usually during dry air conditions, my comforter creates static electricity when I simply pull it off the bed.

Note: I can't move at the speed of light. And guess what happens then? When the air is dry and the room is dark, I see light while I'm pulling the bed spread off the bed. Light is photons, real photons being created between the bed and my bedspread. This is why the gold nucleus flying along in the miles-long machine has photons surrounding it. It's the magnetic lines of force being torn to shreds and creating static electricity, which is magnetic lines of force, just in pieces and no longer attached to an element or molecule.

Off the Cuff—Magnetic Fields Phenomena

One other thought. So scientists are able to make a ball-shaped fuzzy structure that they know is made of possibly trillions of photons, and at its center is an infinitesimally small nucleus of a gold atom. They need very high-intensity magnetic fields that create this phenomenon.

This sounds like the structure some people have observed throughout the centuries called ball lightning (which I noted in the previous chapter). Also, there are many reports that this ball lightning can pass through glass and walls without stopping. If so, it may consist of only one- or two-dimensional mass. One other common factor is, they appear during thunderstorms. Lightning strikes are very high-energy electromagnetic phenomena. Similar to their very high-magnetic-energy machine that makes the photon ball travel near the speed of light.

Similar phenomena under somewhat similar circumstances. Just a thought to add to the fray! There is precedent for the thought that magnetic forces can produce three-dimensional matter such as the electron and the proton. And since we know that an electron added to a proton makes a neutron, we have all the components of an atom.

Constantly I find myself mulling the wide world of physics. After all, there's so much about this mysterious Universe that remains unexplained, as evolutionary scientists will tell you. The question remains: Will self-professed nerds like me, researchers, or people like you, who put to work outside-in thinking, be able to outsmart the evolutionary cosmos, or will the Universe get the better of mankind?

NEW THOUGHTS, OLD UNIVERSE CLUES

- ✓ Two-dimensional Shaylik can bundle up when it's entering a mass like a planet to become two- and even three-dimensional matter.
- ✓ The Breit-Wheeler process hints at the creator of electrons and positrons through the collision of light particles.
- ✓ Magnetic fields can create photons under certain environmental conditions.
- ✓ Earth and the Universe will continue to evolve, whether mankind lives on—or not.

A FINAL WORD

Finally, you are at the end of *The Evolutionary Cosmos*. I've shared my outside-in thinking about the Universe. During the journey, in the beginning, I took you along the road of how the concept and word "Shaylik" came about and deserved to be the focus of this book. Then it was on to an introduction to chirality as a key feature of Shaylik that has a great effect on our environment and biology. But we didn't stop there.

The next stop was the cosmic connection of star to star and how Shaylik affects planets. We got up close to Mars and Ceres—and how Shaylik has an amazing effect on planet environments. And I didn't stop there . . .

It was moving on to gazing at the Universe in a new light, including presenting how Shaylik is a mechanical explanation for gravity. Plus, I shared explanations of dimensionality as another key feature of Shaylik. Not to forget how Shaylik can affect the planets' orbits to create harmony. Then I revealed another key structure of Shaylik, the double-helix. And lastly, the notion that Shaylik may not be totally invisible in X-ray light.

Next it was an inside peek at solar storms. I showed you how Earth's magnetic fields may detect star-to-star Shaylik to how Mercury may be the Rosetta Stone to explain solar flares, which can adversely affect Earth. And I also indicated how Mercury could start solar tsunamis and chain reactions on the Sun. We also pointed out that the stars involved in most significant Earth-striking CMEs are caused by the same star conjunctions as the Mars global dust storms.

After that, it was time to introduce the idea that the planets' positions in relation to the nearby stars are responsible for at least some coronal mass ejections that hit Earth—and can cause unpleasant consequences. I added speculation that Shaylik is like a continuously flowing magnetic river. What's more, there's the idea that some forms of Shaylik, like the double-helix, can be reproduced in quantum steps from the very small to the very large.

Last but not least, I revisited Charles Darwin and Darwin's theory of evolution. Plus, I showed how Shaylik is the basis for Albert Einstein's general and special relativity theories.

OK. Nope, this isn't your basic high school textbook on the Universe. It's a forward-thinking look at the cosmos, past, present, and beyond. So keep an open mind, use this book as your go-to guide when events happen on Earth and in space. Most of all, practice outside-in thinking and draw your own conclusions.

A Few Afterthoughts

My son was also a great help. Many times, we would be discussing a subject and I would experience an epiphany of understanding. Or he would show me something he saw on the internet that brought me to an epiphany.

And last but not least, I need to recognize that the internet in the twenty-first century is a helpful and ever-changing source (like the cosmos). If I asked the right question, I got many hints and leads from other inquiring minds around the globe. I couldn't have done this research without the wonderful tools that were created by others for my use and yours too.

Oh! There was one other key player in this. My brain was an amazing help. If I was confused or unsure about something, I would sleep on it. Many times, the next morning, I'd have an answer. Some ideas made sense; some did not. One thing I came to realize is, my sleeping brain has no morals or social issues to interfere with in coming up with a solution!

A Bonus:

Five Must-Have Cosmos Tools to Use

(Images from theplanetstoday.com with permission from Hayling Graphics Limited (For more information: https://www.theplanetstoday.com/)

Welcome to an online application that provides a quick look at the relationships of the planets and our Sun. This does not predict a conjunction, just the possibility of one. It will allow the user to look at the past, present, and future. It's easy to use and simple to navigate the time. The rectangular box showing the time and date is hidden until the user opens it. Time and date are in the European format, with time, then day, then month, and then year. This is, of course, Earth-related information. The moveable green dots are self-explained upon use by mousing them right or left.

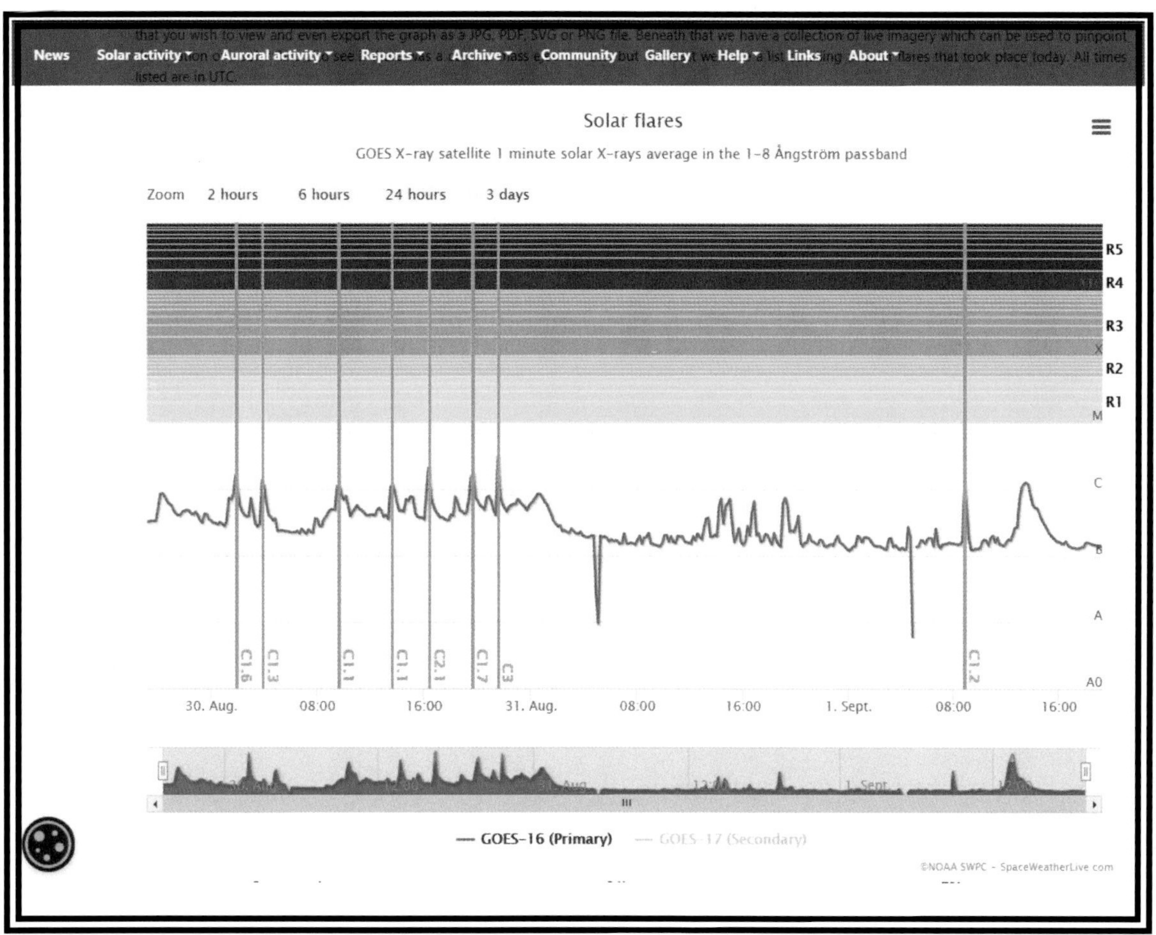

(For more information: https://www.spaceweatherlive.com/en/solar-activity/solar-flares.html)

This site is available online for anyone to use. There are many options as listed in the gray bar. The current option is the GOES satellite data on X-ray energy, among other options on this page. It performs a great service in providing near-real-time information about our Sun and Earth. One key purpose is to warn us of coming coronal mass ejections. Some users need this information so they can take actions to protect Earth-based or satellite-based electronics. The user will need to explore these pages to become familiar with the information available.

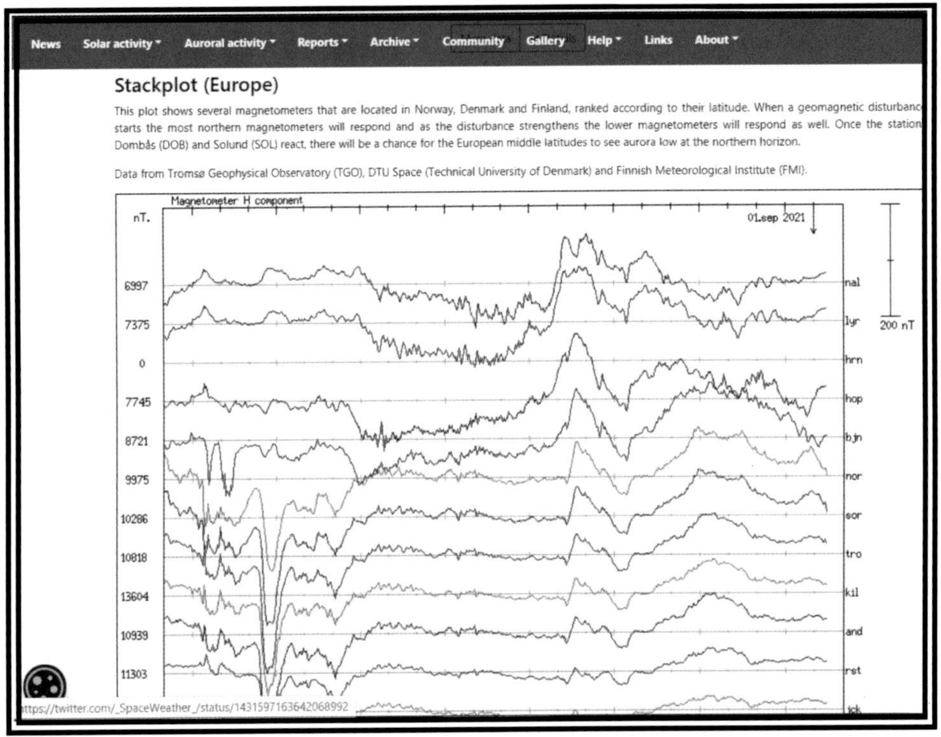

(For more information: Https://www.spaceweatherlive.com/en/auroral-activity/magnetometers.html)

The link is to another page of the spaceweather system. This one shows some of the page data available, in this case the readings from the European magnetometers, which is only one of the options on this page.

There are magnetometers all over Earth, placed at permanent stations. The data from these magnetometers is available in near-real time at the site above. This is data showing the effects of our Sun, our Moon, coronal mass ejections from our Sun, Shaylik from other stars and planets, and God knows what else. Earth is passing through many sources of Shaylik all the time. Some sources are large in their effect. Some are very small. The magnetometer system was designed to let us know what is happening to our planet's magnetic fields so we can anticipate Aurora- and Carrington-like events of serious consequence.

The system is very sensitive and will show the effects of interplanetary and intersolar Shaylik incursions. The intensity of these incursions is very small compared to what our Sun does to Earth's magnetic lines. So there'll be times when the data is lost in the noise of other phenomena. This is in fact the case most of the time. The best time to see these phenomena is when our Sun is at solar minimum because the ambient levels are low.

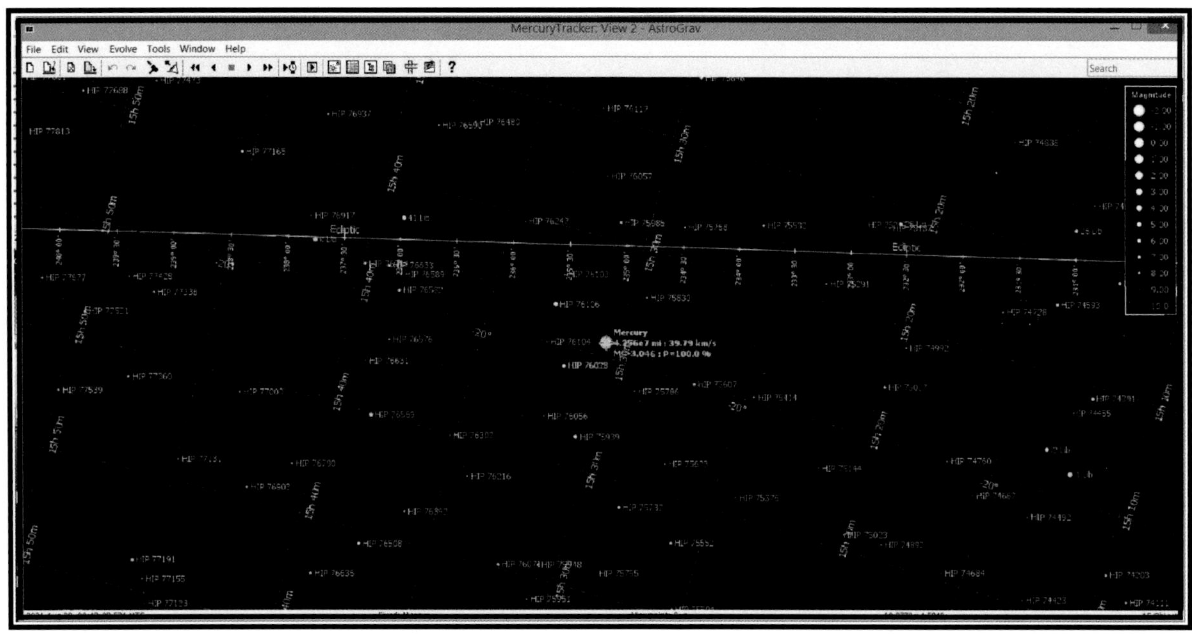

(Images from AstroGrav with permission)
(For more information: http://www.astrograv.co.uk/)

This is a purchased program that can be installed on your computer. It's unique in that this program will give you a view of the sky as it's seen from any planet and our Sun. So it's as if you're on a planet or the Sun using a telescope. Setup for this system can be complicated, and the user needs to use the tutorial supplied with the system. Once the user has a configuration that needs to be saved, all they have to do is give it a unique name and it will then be available from a list of setups to open. I have three setups, for Earth, Mercury, and the Sun. They are all zoomed to easily note one degree of separation, as in the above sample.

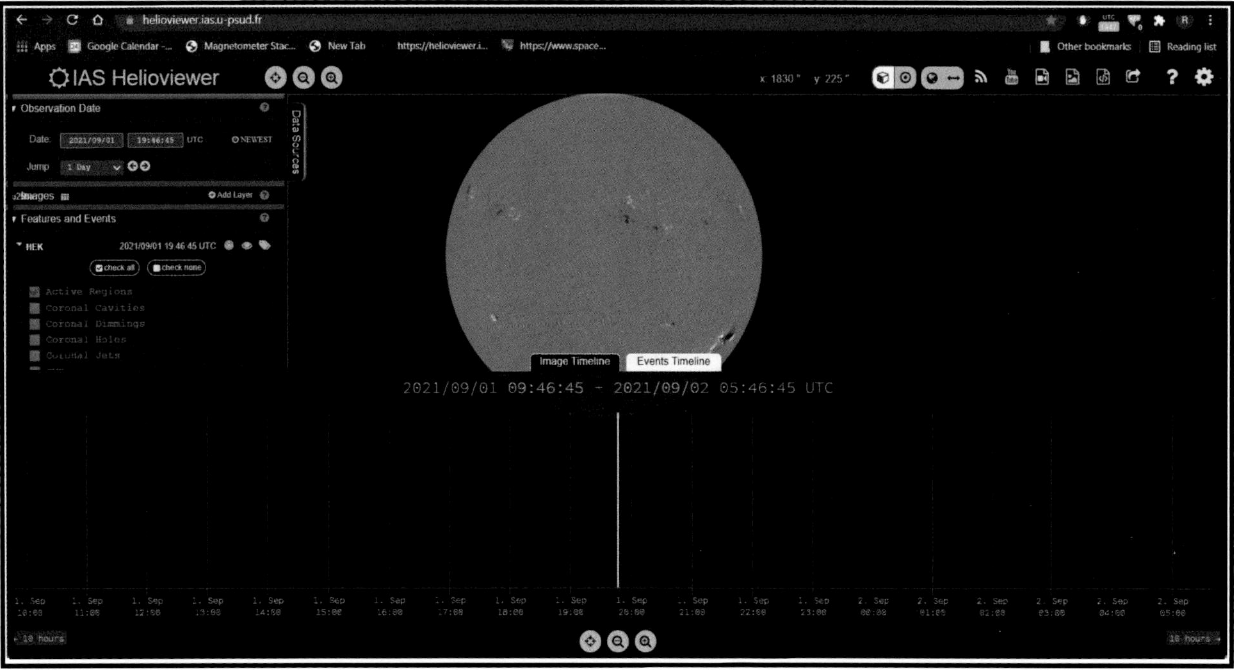

(For more information: https://helioviewer.ias.u-psud.fr/)

This is a wonderful tool for analysis of solar activity over the past two solar cycles. About 22 years of data is available on sunspots, solar flares, active regions, and much more. Any of 20 categories of phenomena can be chosen to view. Also, there are companion pictures of our Sun at the time chosen as viewed from Earth. I can see the exact location of a solar flare if they have located it visually. If not found, it is assumed the flare happened based on X-ray data and is shown at position 0,0, which is the Sun's center at the time. When looking at a single solar-flare event, the HEK data becomes available, which shows the location on the Sun. You may have to zoom in to get this detail.

NOTE: The links and news items provided in the book are subject to change (like our fluctuating Universe). If you find you can't locate the information you're looking for, simply do a Google search on the topic and you'll likely find something similar.

The Cosmos Glossary

Anyon. A word that means a negatively charged ion that is recognized as one- or two-dimensional. In physics, an anyon is a type of quasiparticle that occurs only in two-dimensional systems, with properties much less restricted than the two kinds of standard elementary particles, fermions and bosons.

Ball lightning. An unexplained phenomenon described as luminescent, spherical objects that vary from pea-sized to several meters in diameter. Though usually associated with thunderstorms, ball lightnings are believed to last longer than the split-second flash of a lightning bolt, and are a distinct phenomenon from St. Elmo's fire.

Breit-Wheeler process. The Breit-Wheeler process or Breit-Wheeler theory was founded by two physicists. It's a physical process in which a positron-electron pair is created from the collision of two photons. It's the simplest mechanism by which pure light can be potentially transformed into matter.

Ceres. Dwarf planet Ceres is the biggest object in the asteroid belt between Mars and Jupiter, and it's the only dwarf planet located in the inner Solar System. It was the first member of the asteroid belt to be discovered by Giuseppe Piazzi back in 1801.

Chirality. The intrinsic property of natural materials, including minerals, organic molecules, biological structures, and now magnetic lines of force. It's closely related to the physical and chemical properties of these materials. Also, it plays a role in synthesized materials. There are only two possible chiralities, left-handed spiral or right-handed spiral.

Cosmology. A term that means a branch of metaphysics that deals with the nature of the Universe.

DNA, or deoxyribonucleic acid. It's the hereditary material in humans and almost all other organisms. It's even found in the nucleus of all living cells and is a compact ball structure.

Eddy currents. This isn't how waves in the ocean move, but refers to magnetic drag created in the materials that move by a powerful magnet. Eddy currents are in metal detectors and braking systems in trains and roller coasters, to microwave ovens to heat our foods.

Fibonacci sequence. In mathematics, it refers to the Fibonacci numbers, aka the Fibonacci sequence: each number is the sum of the two preceding ones, starting from 0 and 1. 0+1 is 1, 1+1 is 2, 1+2 is 3, 2+3 is 5, 5+3 is 8, and so on. This physical order is found in organic structures like pine cones.

Fine-structure constant. The fine-structure constant α is of dimension-one or a number and nearly equal to 1/137. It's the measure of the strength of the electromagnetic force that controls how electrically charged particles, such as electrons and photons, work together.

Fractal. A word that means an object of parts, at distinct levels of magnification, that appear geometrically similar to the whole.

Gamma rays. A form of light. The Universe produces a broad range of light, only a fraction of which is visible to our eyes. Other types of nonvisible light include X-rays, ultraviolet light, infrared radiation, and radio waves.

Mass. A term in physics, quantitative measure of inertia, a fundamental property of all matter. It is, in effect, the resistance that a body of matter offers to a change in its speed or position upon the application of a force. The greater the mass of a body, the smaller the change produced by an applied force.

Matter. A term meaning any substance that has mass and takes up space. Basically, it's anything that can be touched or affected by other matter.

Neutrino. A word that means a particle. It's one of the so-called fundamental particles, which means it isn't made of any smaller pieces, at least that we know of. Neutrinos are members of the same group as the most famous fundamental particle, the electron.

Quantum mechanics. It's the branch of physics that deals with the behavior of matter and light on a subatomic and atomic level. It explains atoms and molecules and their fundamental particles, like protons, neutrons, electrons, gluons, anyons, and quarks.

Shaylik. A made-up word that describes the important building blocks of creation. It's the fundamental unit making up all there is or will be in our Universe.

Vortices. In fluid dynamics, a vortex (plural "vortices"/"vortexes") is a region in a fluid in which the flow revolves around an axis line, which may be straight or curved. Vortices form in stirred fluids, and may be observed in the winds surrounding a tornado or dust devil. Vortices are a component of turbulent flow.

X-lightning. This word denotes lightning that is only manifested by its X-ray emissions.

X-ray, an X-ray. A term to describe X-radiation, which is a penetrating form of high-energy electromagnetic radiation. A form of light.

NOTES

CHAPTER 1:
THE CREATION OF "SHAYLIK"

1. https://iai.tv/articles/why-physics-has-made-no-progress-in-50-years-auid-1292;
2. https://en.wikipedia.org/wiki/List_of_unsolved_problems_in_physics
3. "Introduction to Iamblichus Exhortation to Philosophy." Retrieved December 2021.
4. "Inverse Square Law, Gravity," HyperPhysics Mechanics. http://hyperphysics.phy-astr.gsu.edu/hbase/Forces/isq.html.

CHAPTER 4:
ICE VOLCANOES AT CERES

1. "What Is an Ice Volcano?" HowStuffWorks. https://www.howstuffworks.com/search.php?terms=ice-volcano.

CHAPTER 5:
A MODERN MEANING OF GRAVITY

1. "Diurnal variation," The Free Dictionary, By Farlex. www.thefreedictionary.com/diurnal+variation.

CHAPTER 6:
DISCOVERING DIMENSIONS TO ANYONS

1. Ornes, Stephen. "Physicists Prove Anyons Exist, a Third Type of Particle in the Universe." December 12, 2020, 5:00 p.m.. https://www.discovermagazine.com/the-sciences/physicists-prove-anyons-exist-a-third-type-of-particle-in-the-universe.

2. Girvin, Steven M, "Anyons Superconduct, But Do Superconductors Have Anyons?" *SCIENCE* • 4 Sep 1992 • Vol 257, Issue 5075 • pp. 1354-1355 • DOI: 10.1126/science.257.5075.1354 https://www.science.org/doi/epdf/10.1126/science.257.5075.1354.

CHAPTER 8:
A NEW LOOK AT X-RAYS

1. GLEAMoscope. View the GLEAM survey and the sky at various wavelengths. https://gleamoscope.icrar.org/gleamoscope/trunk/src/.

2. "X-ray Astronomy," National Aeronautics and Space Administration Goddard Space Flight Center. https://imagine.gsfc.nasa.gov/science/toolbox/xray_astronomy1.html

CHAPTER 9:
EARTH'S MAGNETIC FORCE

1. Agustina, "10 Facts about Magnetic Fields," Less Known Facts, April 17, 2017. https://lessknownfacts.com/10-facts-about-magnetic-fields/.
2. SpaceWeatherLive.com, Real-time auroral and solar activity. Retrieved December 2021. https://www.spaceweatherlive.com/en/auroral-activity/magnetometers.html.

CHAPTER 10:
SAY HELLO TO PLANET MERCURY

1. Graphic Image. https://spaceplace.nasa.gov/solar-activity/en/solar-activity3.en.gif.
2. Wall, Mike, "Huge Magnetic Bubbles May Churn at Solar System's Edge," Space.com. June 9, 2011. https://www.space.com/11912-nasa-voyager-solar-system-magnetic-bubbles.html.

CHAPTER 11:
SOLAR SUPERSTORMS

1. Francis, Matthew, "Why is the Sun's corona so hot? Two words: Solar tornadoes." Ars Technica, *Condé Nast.* June 27, 2012. https://arstechnica.com/science/2012/06/why-is-the-suns-corona-so-hot-two-words-solar-tornadoes/.
2. "Raining Loops on the Sun," NASA Video. May 22, 2013. https://www.youtube.com/watch?v=1ZxrVlegJ28&ab.
3. Shere, Jeremy, "What Are Solar Tornadoes?" Indiana Public Media. Retrieved December 2021.

CHAPTER 12:
ELECTRIFYING THE EARTH

1. From AstroGrav Carrington and 2017 coronal mass ejection moment details. Retrieved December 2021.
2. "What was the Carrington Event?" SciJinks, *NOAA*. https://scijinks.gov/what-was-the-carrington-event/. Retrieved December 2021.
3. Strain, Daniel, "A 1972 solar storm triggered a Vietnam War mystery," *CU Boulder Today.* November 12, 2018. https://www.colorado.edu/today/2018/11/12/1972-solar-storm-triggered-vietnam-war-mystery.
4. "August 1972 solar storm," Wikipedia. Research retrieved December 2021. https://en.wikipedia.org/wiki/August_1972_solar_storm
5. Ibid.

6. Phillips, Tony, Dr., "Near Miss: The Solar Superstorm of July 2012," NASA Science, *NASA*. https://www.spacedaily.com/reports/Near_Miss_The_Solar_Superstorm_of_July_2012_999.html
7. "List of solar storms," Wikipedia. Research retrieved December 2021. https://en.wikipedia.org/wiki/List_of_solar_storms
8. "M1.6 solar flare, G2 storm watch," SpaceWeatherLive. October 10, 2021, 15:17 UTC. https://www.spaceweatherlive.com/en/news/view/443/20211010-m1-6-solar-flare-g2-storm-watch.html.

CHAPTER 14:
FRACTAL EXPLAINED

1. "Fractals & the Fractal Dimension," Fractal Facts for Kids KidzSearch.com, Vanderbilt.edu. Retrieved 2011-10-28.
2. Mandelbrot, Benoît B. (1983). *The Fractal Geometry of Nature*. Macmillan. ISBN 978-0-7167-1186-5.

CHAPTER 15:
DARWIN'S THEORY OF EVOLUTION

1. "Evaluation of Charles Darwin's Theory of Evolution." (2019, February 12). GradesFixer. Retrieved January 1, 2022, from https://gradesfixer.com/free-essay-examples/evaluation-of-charles-darwins-theory-of-evolution/.

CHAPTER 16:
RETHINKING EINSTEIN'S GRAVITY THEORY

1. "Time dilation," Einstein Online, Dictionary, *Max Planck Institute for Gravitational Physics, Potsdam.* https://www.einstein-online.info/en/explandict/time-dilation/. Retrieved December 2021.

CHAPTER 17:
THE EVOLUTION OF OUR UNIVERSE

1. Conover, Emily, "Light is caught making matter," *Science News*. Retrieved December 2021. https://www.sciencenews.org/article/colliding-photons-matter-particle-physics.
2. Starr, Michelle, "Physicists Detect Strongest Evidence Yet of Matter Generated by Collisions of Light," Science Alert. August 10, 2021. https://www.sciencealert.com/physicists-claim-they-ve-finally-observed-matter-being-made-out-of-colliding-light.

Epilogue

Mankind has named many phenomena, and defined what each did. Diseases have been named and described. For some, we have found the cause, and for some of those, we discovered a cure. Medical science has been one of the greatest benefits to mankind there is. An equally great benefit to mankind was due to the work of one man. Norman Borlaug showed the world how to grow high-quality, high-yield wheat crops that have fed the world since and led the way to improve all our food production.

Now that we know not only what gravity does but also what it is, we can learn more about how to live with it. Overall, gravity has been kind to our Universe, and mankind in particular. There are some disasters that happen due to gravity, and some are our fault for not understanding them. What great things may come of this knowledge are to be determined. One thing I do know is, it's up to us to keep this knowledge beneficial to mankind.

One more comment is necessary. My son and I have only considered those phenomena that are stable in our Universe. This is logical because the phenomena we were trying to understand were themselves stable. Gravity, for instance, is about as stable as anything gets. So what it is all about, what causes it, must also be very stable in our Universe. We didn't allow much thinking about things, no matter how real they may be, that could not last more than minutes in the cosmos. This approach saved us a lot of time by not taking the wrong turn on the highway to the Universe.